Modern Statistics:

An Introduction

by Don McNeil and Jenny Middledorp

Pearson Originals is an imprint of Pearson Australia, a division of Pearson Australia Group Pty Ltd. It has been established to provide academics throughout Australia and New Zealand with fast and efficient access to the printing, warehousing and distribution services of Australia's leading educational publisher, ensuring a smooth supply channel to your campus bookseller.

For more information about the **Pearson Originals** service, contact the Editorial Department, Pearson Australia, Unit 4, Level 3, 14 Aquatic Drive, Frenchs Forest, New South Wales, 2086. Telephone: 02 9454 2200.

Photo Credits
Front cover (from left to right): © Holimage, © Paul Wright, © Holimage, © Paul Wright, back cover: © Brett Cornish, chapter opening pages: © Paul Wright.

Managing Editor: Jill Gillies
Learning Solutions Consultant: Jennifer Ellis
Senior Project Manager: Lisa D'Cruz
Production Controller: Caroline Stewart
Indexed by Sue Hill, Hill Associates

Typeset by Midland Typesetters, Australia

ISBN 9781486007011

Pearson Australia
Unit 4, Level 3
14 Aquatic Drive
Frenchs Forest NSW 2086
www.pearson.com.au

Acknowledgements

Ours thanks are extended to many members, both past and present, of the Statistics department at Macquarie University for their support in producing this book.

Table of Contents

INTRODUCTION

CHAPTER 1

1 Introduction

For most of you, Statistics is not a new area of study, and you will have encountered statistical ideas at school and in everyday life. You have probably developed some opinions about the subject. As you learn more, you will find that these opinions will change.

Statistics is changing rapidly, perhaps more rapidly than any other area of study. This is largely because of the dynamic expansion of information technology over the last 50 years, and especially over the last 20 years. The changes in the next 10 years are likely to be just as rapid and dramatic.

The new technology has changed the practice of Statistics. In the past it was difficult to view your data, and analysis was done by hand or at best with a cumbersome calculator. Nowadays data are stored on computer disk, visualised in multiple ways with the aid of fast and powerful colour graphics terminals, and analysed in a flash just by mouse-clicking a few buttons. The results are easily presented in full colour with informative graphical illustrations using modern windows technology.

The new technology is also changing the way we learn Statistics. In the past, statistical textbooks were loaded with mathematical formulas and tables and not so many graphs, and theory was emphasised at the expense of practical applications.

The new technology has made redundant many of the statistical methods students were forced to learn in past decades. It is no longer necessary for statisticians to calculate for themselves quantities such as means, standard deviations, t statistics and regression coefficients. Once the data are stored in a computer file, software is available, not only to do such routine calculations, but also to produce appropriate graphs and even to undertake appropriate analyses.

So what is the modern student of Statistics expected to know?

The answer is that the new technology has given statisticians a much richer and more important role in the conduct of research, so students in Statistics now need to learn the methods for undertaking research studies. Now that powerful knowledge-based statistical software is taking over most of the routine tasks statisticians used to carry out, students in Statistics also need to learn how to use these software tools effectively.

And since much of the data used to answer research questions needs to be stored and managed in highly structured databases, modern students of Statistics need to learn modern database technology as well.

Perhaps most important of all, statisticians need to be able to present data and results of studies effectively. With the advancement of windows technology, presentation of data and results now demands effective graphical methods for data analysis.

It is in the use of graphs to present statistical results that the change in Statistics has been most noticeable. So in this book graphical methods play a leading role. In the past graphs were not easy to produce routinely, and consequently were used sparingly in statistical texts, often with the aim of catching the reader's attention in much the same way that comic strips are included in newspapers. Nowadays the situation has changed almost to the extent that graphs are used routinely to convey the main results and explanatory text is added to assist interpretation.

Statistics is a broad field that includes the design of research studies, the analysis of data, and the interpretation of the results of studies, raising further questions leading to the need for new studies to be undertaken. These topics are considered in some detail in the following sections.

Designing a study

Statistical methods are used in the process of doing studies to answer questions of interest. So to properly understand the role of Statistics it is necessary to understand what a research study entails.

A research study attempts to answer a question of interest, involving a target population, using a specified set of research methods.

The question of interest

The question of interest in a study is usually concerned with determining either a level or an association.

For example, you might wish to know what is the mean level of bacterial contamination in Sydney's drinking water. Or the proportion of Australian residents currently in favour of the nation becoming a republic, or the incidence of major volcanic eruptions in Java, or the risk that a student will fail in a university course. Each of these questions involves the measurement of a level for some quantity of interest.

An association involves two quantities. You may wish to know the extent of the association between drinking alcohol and developing stomach cancer, or between the weight of a foetus and its gestational age. Or between the engine capacity of a motor vehicle and its fuel consumption, or between the nature of a teaching method (traditional or problem-based) and the academic performance of the students.

Having stated *what* is the question of interest, you need to say *who* or *what* are the subjects and explain *why* it is of interest to study them. A study should be interesting and important. Doing a dull and boring study is a waste of everybody's time.

The target population

The target population for a study is the set of all subjects to which the results of the study may apply. Usually it is not possible to take measurements from all members of this population, because it would be too costly or time-consuming, so the study is based on a sample from the target population.

For example, if you wish to determine the level of bacteria in the drinking water supplied to a city, the target population could comprise every millilitre of water supplied to the residents of the city during a given period. In the determination of the incidence of major volcanic eruptions in Java, the target population is simply all the volcanic peaks in Java. For a study involving the determination of the association between alcohol consumption and stomach cancer disease, the target population could be a specified demographic group, such as Filipino men aged between 20 and 60.

So the words *who* (or *what*), *where*, and *when* are used to describe the target population for a research study.

The research methods

The methods for conducting a study need to be specified in advance. These include the following components.

a. the number of subjects in the sample (that is, the sample *size*), and the method of selecting these subjects from the target population;

b. the characteristics to be measured, and the instruments used to collect these data;

c. the statistical methods used for analysing the data; and

d. the timing and the method for reporting the conclusions from the study.

For example, the level of bacteria in a city's water supply could be investigated as follows:

a. collect samples in one-litre bottles from 300 randomly selected households in the city;

b. take these bottles to a central laboratory and measure the concentration of *e coli* contamination in each sample;

c. compute the mean of the 300 concentrations and give a measure of error (to be explained in detail in Chapter 6) for the corresponding population mean, and

d. enter these results (including the details of the locations of the sampled households and the times at which the measurements were taken) into a database maintained by the Department of Health.

Similarly, you could study the association between alcohol consumption and stomach tumours as follows. First, use hospital records to find the next 100 men in the Philippines aged between 20 and 60 who are diagnosed with this disease. Then select another 100 similar men who are free of the disease and compare the alcohol drinking history of the two groups. Finally, publish the results in a medical journal.

Types of studies

Preliminary research is needed to refine the research question. First, you need to search the literature to see what relevant studies have already been undertaken. It is possible that the question has been studied in other places, perhaps under different conditions. Such preliminary investigation will determine how much further data need to be collected.

Deductive studies

Most studies require data collection, but this is not essential. A study could be purely *deductive*, based on logic, starting with known facts or assumptions and arriving at a conclusion. Mathematicians and theoretical physicists do research by starting with known scientific principles or axioms, and using derivations based on algebra and calculus. So a mathematical proof is an example of a deductive study. But deduction needn't involve mathematical formulas, as the following example illustrates.

Patients in the intensive care unit of a certain hospital ward are dying inexplicably and regularly. Most of the deaths occur on Friday evenings. Could the spate of deaths be attributed to chance? The hospital director calls for a study to be undertaken to explain why these patients are dying.

The question could be addressed by undertaking an observational study, involving the collection of detailed information about the patients over a period of time, and carrying out extensive post-mortem examinations to determine the causes of death.

In fact the deaths have a very simple explanation. The hospital cleaner, who works every Friday evening, except for long weekends when she works on Thursday evening instead, finding no available power socket for her vacuum cleaner, briefly disconnects the power supply for a patient's life-support system, unaware of the consequences of her action. A good statistical detective could possibly deduce this finding without further data collection.

Knowledge of appropriate theory can also reduce the need for data collection. Suppose you wish to undertake a study involving persons in a region within a specific age group (say 20-24) and you need to know the size of the target population. Suppose further that the only information you have is the total population of the region (N, say). Now you could get an estimate of the number of 20-24 year-olds by visiting every household in a manageable area, computing the proportion of persons aged 20-24 in this sample, and multiplying this proportion by N. But you can get a more accurate estimate, without collecting any data, simply by using some basic demographic theory, namely that, in a stable population, the proportions of persons in each 5-year age group are similar. So if the life expectancy is 70, say, there will be close to 7% of the population aged 20-24.

Simulation studies

Computer simulation is another option than can avoid the need to collect data. For example, it was claimed by some scientists that the French nuclear tests carried out in the South Pacific in 1995 and 1996 were not necessary because the information gained from them could have been obtained using simulation based on computer models.

So the first question to be answered before undertaking a study is to determine whether any data need to be collected at all (and if so, how much).

Empirical studies

If a study involves data collection it is called *inductive*, because the data in the sample are used to induce the characteristics of the target population. Unlike a deductive study, where the conclusion may be stated with certainty provided sound logic is used, an inductive study can only provide conclusions that are approximately correct. For example, if you wish to determine the proportion in a country's population favouring a change of government, based on a poll taken from a sample of residents. You cannot determine this proportion exactly: the best you can do is get an estimate which is approximately correct, perhaps to within 5%.

Inductive studies based on data collection are also called *empirical* studies, and may be quantitative or qualitative.

Quantitative and qualitative studies

A *quantitative* study involves structured data collection. In this case the same characteristics are measured from all subjects in the study, using a protocol or set of guidelines that are specified in advance. The method of data collection in a quantitative study is often a questionnaire, or an instrument that automatically records the sample of measurements. In contrast, a *qualitative* study is relatively unstructured. It may involve open-ended questioning of subjects, and may be opportunistic in the sense that the answer given to one question determines the next question. A qualitative study may precede a more formal quantitative study, with the aim of getting sufficient information to know what measurements to record in a quantitative study. For example, an investigator wishing to compare a new program for improving the reproductive health of mothers in an aboriginal community in outback Australia would be advised to spend some time in the community getting to know the people before embarking on the fully-fledged study.

Descriptive and comparative studies

A quantitative study may be descriptive or comparative. A *descriptive* study is aimed at measuring the level or prevalence of some outcome. For example, you may wish to know the proportion of small businesses in Singapore that go bankrupt in a year, or the mean wing span of sea eagles in the northern beach suburbs of Sydney, or the average level of atmospheric carbon monoxide in Bangkok.

A *comparative* study is concerned with measuring or testing the existence of an association between two variables of interest. For example, a medical scientist may wish to know whether a relation exists between the amount of beer consumed by men and their risk of developing stomach tumours. An economist may be interested in the association between the unemployment rate and the inflation rate in an economy.

A descriptive study is concerned with a single outcome variable, whereas a comparative study measures two characteristics. One characteristic is called the *outcome* and the other is called the *determinant.* The determinant may be a cause of the outcome, but the outcome cannot cause the determinant. Thus in the study involving the beer-drinking men, the outcome is the disease and the determinant is the amount of beer consumed.

Note that outcomes and determinants are not always clearly defined. In a study of the association between inflation and unemployment, which is which? Two characteristics may each be caused by one or more other variables. The question of what causes what needs be considered. For example, a TV advertisement has a doctor from a medical research institute saying "people with pets tend to get fewer minor diseases, like colds and flu". This suggests to the viewer that if you get a pet you will become healthier. Of course it is more likely that the causal association works the other way, that is, healthy people own pets to a greater extent than unhealthy people. You could speculate on the reasons.

Experimental and observational studies

Studies may be experimental or observational. An *experimental* study is one in which the investigator has some control over a determinant. Suppose, for example, you wish to investigate a possible association between beer consumption and stomach tumour incidence. You could take a sample of laboratory mice and divide them into two groups, force one group to drink beer and deprive the other, and observe each group to see how many mice developed the tumours over a period of time.

Observational studies do not involve any control by the investigator over the determinant, and are thus in a sense less rigorous than experimental studies. However, experimental studies require interventions that may be costly, and are not always feasible or ethical, particularly if human subjects are involved.

Case-control, cohort, and cross-sectional studies

Controlling the selection of the subjects may increase the efficiency of a study. For example, you could study a possible association between beer drinking and stomach tumour incidence by taking a sample of all men being treated in hospitals, asking them about their beer drinking habits, and comparing the proportions with stomach tumours between the beer drinkers and the abstainers. However, since these tumours are rare, the sample would contain a very small proportion (probably less than one percent) of men with the outcome. A more efficient study would include all the hospital patients diagnosed with stomach tumours together with a similar number of selected subjects free of this disease. For example, each patient with a tumour could be matched with a patient from a nearby hospital bed. Thus, instead of having a study with, say, 5000 subjects (100 with a tumour and 4900 without one), the study would need only 200 subjects (100 in each outcome group).

This type of study is called a *case-control* study, because the cases (the subjects with the outcome) are compared with *controls* (subjects without the outcome, selected from the target population). Note that different proportions of cases and controls are selected. Usually, all the available cases from the target population are selected, but only a small proportion of controls is selected.

Another type of study is a *cohort* study. A cohort study is similar to a case-control study, but subjects are selected according to their determinant status rather than their outcome status. Suppose, for example, the police department would like to know if women officers are better than their male colleagues in dissuading suicidal people from jumping off buildings. A cohort study of this question could involve allocating equal numbers of men and women police officers to this task in future, and comparing the success rates of the two sexes after observing a specified number of suicide attempts.

A third type of comparative study is a *cross-sectional* study. In a cross-sectional study, there is no differential selection of subjects, either by outcome or by determinant. In a cross-sectional study you simply select subjects from the target population, without taking into account the outcome or the determinant. For example, suppose the question is ‘do 10 year old girls do better than 10 year old boys at Maths?' You could do a cross-sectional study by taking a sample of 10 year-olds, giving them a Maths test, and comparing the boys' results with those of the girls.

Choosing the sample

The statistician's objective is to get information about certain characteristics of a target population, using a sample.

The target population comprises all relevant subjects of interest. For example, you might be interested in all European professional soccer players, a population of several thousand young men.

The sample is a manageable subset of the target population, selected to make the study feasible. For example, you could select two players at random from a team in each of ten European cities.

The measurements taken - the data - are called *variables*. For example, one variable could be the age of the subjects, another their salary level, and another their opinion on some question, expressed in terms of the extent to which they agree or disagree with a statement on a questionnaire.

In this section we discuss some properties of samples: their selection, how well they represent the target population and their accuracy for determining population characteristics.

Random sampling

The sample should be a representative sample from the population. This means that it should not be biased, or atypical. In the sample of soccer players, choosing the best teams from the larger cities could give rise to a biased sample, because these players are likely to receive larger salaries than players from smaller localities. It would be better to have a list of all the professional soccer teams playing in competitions in Europe, and then choose ten teams at random from this list, making sure no more than one team is chosen from each city.

A simple random sample of size n can be defined as one in which each sample of size n has an equal chance of being selected from the target population.

Does the method of selection have to be random? Not necessarily, but choosing a sample using random selection uses a chance mechanism to guard against systematic biases than can occur. Random selection does not guarantee that the sample is representative, because random events can give rise to unlikely sequences (such as the captain of a sporting team losing the toss eight times in a row), but if the sample is sufficiently large, such biases tend to average out.

To see why this is so, consider an alternative sampling procedure. When market researchers in the US first started sampling opinions from potential customers in the 1930s, the rule was to make sure that their samples contained subjects from all the important groups, such as men and women, people of different ages, blacks and whites, people with different incomes, occupations, etc. So the pollsters made sure they selected persons from each subgroup (e.g., black male blue-collar workers), without concern for the method of selection. But this created another problem: the persons polled tended to be the most conspicuous or easy to find, with the result that a lot of people (such as people with disabilities, shift workers, and introverts) were overlooked, and the samples were thus unrepresentative of the target population. To overcome this problem, market researchers now use a method called stratified randomisation, in which the subjects are selected at random from each subgroup. This method ensures that the sample gives adequate coverage of the target population, as well as being representative.

How is a random sample selected? Suppose that the population contains m subjects and the sample is to be of size n. Conceptually, the simplest method is to make a list of all the subjects in the population, giving them numbers from 1 to m (where m is the total population size), and then choosing a set of n numbers from the list $(1, 2, 3, \ldots, m)$ using a computer random number generator. For example, if m is 2000 and n is 10, the computer might give the numbers (1216, 53, 876, 1913, 665, 901, 435, 1265, 992, 1552). Note that this method requires that all the members of the target population be listed, like subscribers in a telephone directory, or voters on an electoral roll.

In practice there are more convenient methods of choosing a representative sample. For the telephone subscribers, it would be simpler to choose random pages from the directory and then to select one subscriber from each selected page. (For example, if the directory has 2000 pages and you require a sample of 10 subscribers, you could use the computer to select the pages as 1216, 53, 876, 1913, 665, 901, 435, 1265, 992, 1552. Then select a subscriber from each page, such as the tenth subscriber listed on each page.)

The main requirement when selecting a random sample is to ensure that each sample of size n has the same chance of being selected. In practice this is often not possible or may be very difficult to achieve because of time or financial constraints, for example. However, it is important that, to ensure the sample is not biased, a representative sample is selected for any statistical analysis.

If the population is classified into subgroups, like men and women, old and young, drinkers and abstainers, etc., the chance of selection may differ from one subgroup to another, but it still should be the same for each member of the subgroup. Suppose, for example, your research question is to compare the attitudes towards fees of university students from different States of Australia, and resources limit your sample size to 120. In this case you may prefer to use random selection to get 20 students from each of the 6 States, even though this method will give a student from Tasmania a much greater chance of selection than a student from New South Wales. The important requirement is that each student from a given State should have the same chance of selection as any other student from the same State.

Exercise 1.1:

Consider the persons in your class.

a. *What population do they represent? In other words, from what population could they be considered to comprise a representative sample?*
b. *Consider the ages of your classmates. Is this sample of ages representative of the age distribution of the Australian population? Explain.*
c. *Consider the ethnic composition of your classmates. Is the class representative of the ethnic composition of the Australian population? Explain.*
d. *Consider the proportion of women among your classmates. Is this proportion an unbiased estimate of the proportion of women in the Australian population? Explain.*

Bias

Bias is a kind of distortion that makes a study provide wrong results. It can arise in three ways as follows: (1) through poor measurement; (2) due to the sample not being representative of the target population; and (3) by failing to take into account an important factor when doing the statistical analysis.

The first kind of bias is called *measurement* bias. It arises when an instrument used to obtain the data gives consistently wrong values in a fixed direction. Examples include bathroom scales that say you are 2 kilograms lighter than you really are, and teachers who inflate the grades of their students in an attempt to gain their affection. This kind of bias can also arise when subjects are asked to recall their experiences and they forget events that are painful to remember or occurred a long time ago.

The second kind of bias is called *selection* bias. It can arise for a variety of reasons when the sample is not representative of the population. But even if the sample selection is satisfactory, further problems can occur at the data collection stage, giving rise to response bias.

Response bias can arise when the available data are reduced because selected subjects fail to provide data. It arises frequently in studies where questionnaires are mailed to respondents. Usually people who fail to respond to such questionnaires are atypical.

A celebrated historical example of response bias in a sample is the poll of prospective voters in the US presidential election in 1936 undertaken by the Literary Digest magazine. Of a sample of 10 million persons invited to return a mailed questionnaire indicating their voting preferences, 2.4 million responded, of whom 57% said they would vote for the Republican candidate (Landon). As it turned out, the Democrat Roosevelt won the election in a landslide (62% to 38%). The poll got it wrong because most of the subscribers to the Literary Digest were atypical of the population (richer and thus more likely to vote Republican) and therefore were a biased sample.

In practice a response rate of at least 75% is needed for a study to be considered free of response bias.

Response bias can also arise when data are not recorded for all variables of interest. This can create a problem in the statistical analysis, because the size of the sample may depend on which variables are included in the analysis.

The third kind of bias in a study is called *confounding* bias, and arises in comparative studies where the association between a determinant and an outcome is distorted by the presence of another determinant. For example, the death rate in suburbs of Sydney is highly correlated with income: the suburbs with higher per capita incomes tend to have higher per capita death rates. But this result is biased because the more affluent suburbs have older residents on average, and older people have higher mortality. So we say that the association between income and mortality is 'confounded by age'.

Accuracy

Recall that the objective of Statistics is to get information about characteristics of the target population. Ideally, the information obtained should be accurate.

If a sample is too small, it will not provide sufficiently accurate information about the population. It is also important for the sample to give some coverage of the different components in the population, and if the sample is too small it may not be possible to do this. Suppose, for example, you wish to sample the opinions of citizens with respect to some issue of major national political interest (such as when and how the nation could become a republic). Then the sample should represent the different groups in the population, such as farmers, aborigines, migrants, conservatives, progressives, men, women, and people of different ages, religions, income brackets, occupations, etc. A sample of several hundred would be needed.

Nor should the sample be too large, because large samples need more resources. The sample should be large enough to gain adequate information, but not larger. If a sample is too large, resources are wasted, which could better be used in a follow-up study. A good study usually answers the question of interest, at least partially, and then raises further questions, justifying further studies.

The accuracy of a sample in determining a population characteristic does not depend on the size of the population. This fact may seem surprising at first. You might expect that the bigger the population, the bigger the sample needed to achieve a specified accuracy.

In fact the accuracy of a sample for determining a population characteristic depends only on two factors, namely (1) the size of the sample, and (2) the extent of the variability of the characteristic in the population. To see why this is so, consider the following example.

Suppose you wish to estimate the average weight of grains of sand at a specified beach. The size of this population is effectively infinite. But if you took a representative sample of, say, 1000 grains of sand and weighed this sample, you could get a very accurate determination, probably within 1% of the population mean. It does not depend on the total amount of sand in the beach. As long as your sample is representative of the population, a sample size of 1000 should be more than enough to tell you quite accurately what is the average weight of a single grain.

Would a sample of 100 grains be large enough? Possibly. It depends on the variation in the weights of the individual grains. If the grains were all exactly the same size, you would only need a sample of 1! But if they varied a lot, a sample of 100 might not be large enough. So the range of the measurements in the population has a significant effect on the accuracy with which you can determine a population characteristic.

Note that this means that if you are doing a comparative study to compare two populations, it is usually better to take samples of equal size from each population, even if the population sizes are markedly different. This principle was used in examples given earlier in the chapter. You saw how (a) a case-control study involves matching cases with an equal number of controls, and (b) comparing characteristics of populations in different States involves selecting the same number of subjects from each State, even though the population sizes are quite different. However, there could be other situations in which it would be better to select sample sizes in the same ratio as they occur in the different states.

Probable error

It is useful to define the accuracy of a sample for determining a population characteristic in terms of the *probable error* between the sample measurement and the population characteristic. The probable error is a measure of how close you expect the sample measurement to be to the population characteristic, on average. This will be outlined in more detail in Chapter 3.

Suppose, for example, that you wish to determine the average cholesterol level of adult Australians, and a representative sample of 10 gives a mean value of 4.9 millimoles per litre. If the probable error based on this sample is 0.2, you may conclude that the population mean is probably between 4.7 and 5.1 millimoles per litre. This statement can be made more precise by specifying the probability as an exact quantity (e.g., 50% probable, 90% probable, etc.), a concept we discuss further in Chapter 6.

The accuracy depends on the variability of the measurements in the population. This means that a more accurate determination can be made in a population with smaller variability.

Determining a proportion

If the population characteristic of interest is a proportion it is also useful to give a measure of error for the population proportion.

As an illustration, suppose you wish to take an opinion poll to see how many voters support the 'Don Juan Nation Party' in Spain. Initial surveys suggest that the level of support is around 20%, with a margin of error of 1%. This means that the level of support is probably between 19% and 21%. Again the statement can be made more precise by specifying the probability as an exact quantity.

Presenting the results

It is important to know how to present the results of a study.

Even if a study has adequately answered an interesting research question, it will have been a waste of time if the results are not reported effectively.

In a study of the effect of new Sydney airport development in 1994 on the noise suffered by residents living in the vicinity of the runways, a report presented contour maps showing the predicted levels. However these maps failed to give any indication of the likely variation in noise levels around their predicted values, and thus caused a political storm among residents because property valuations had failed to take this variation into account.

One of the best methods for presenting the results of a study is to use appropriate graphs. Chapter 2 covers some of the basic methods for graphing data, including histograms, pie charts and bar charts, box plots, and cumulative distributions.

In more complex situations, a high-resolution graph can depict a large amount of data very concisely, particularly if the graph has an appropriate explanation. As John Tukey said in his 'sunset salvo' (1986), "a picture may be worth 10,000 words, but it may take 100 words to make it so".

In any case the key to making a good graph is to have good computer software, and the data need to be properly stored and managed.

As well as providing appropriate graphical displays it will also be necessary to provide numerical summaries of the data. For conciseness this will usually mean presenting the results in a table. Of course, the numerical summaries will complement the graphical displays provided.

Parameters and Statistics

The numerical summaries you obtain are called statistics because they arise from the sample. They are estimates of the population parameters.

Managing data

Data management entails effective measurement, collection, maintenance, analysis, and reporting of the data needed in a study.

There are several aspects to measurement. You need to decide *what* to measure, *who* are the subjects providing the data for the study, *how* to make the measurements, *how many* measurements to take and *when* and *where* to take them. It is also important to understand the nature of the errors that can arise in data collection.

Some measurements, like a person's age, weight or gender, or the depth of snow on your favourite ski run, are relatively easy to measure. Many measurements of general interest, like sporting records and population changes, are readily available from libraries.

Some quantities are difficult to measure, requiring complex equipment or sensitive instruments. Measuring attitudes of subjects in an opinion survey is often quite a challenge. So is measuring levels of water and air pollution in an environmental study.

Data management has increased in importance with the development of computer technology, which makes it feasible to undertake studies involving the collection of massive data sets from millions of subjects. Nowadays remote sensing instruments collect time series data automatically, and the challenge is to summarise the data in a meaningful way, separating the information from the noise.

While computers have given statisticians the power to manage huge amounts of data, they have brought their own problems. Statisticians need to learn concepts in information systems if they are to effectively control their data, and failure to master the relevant computer software can lead to serious problems including the corruption of data records. For example, studies involving hospital records of patients depend for their funding on proper standards of data management and data quality control.

Data management in research study teams is similar to funds management in investment companies. The Statistician is responsible for preserving the integrity of the database, just as a financial systems manager must account for the flow of funds. Both provide a service to the other members of the team, and should be in a position to respond to queries about the information in the database, answering routine requests for reports.

Finally, the reporting of results from studies has become a required skill for the modern statistician. As in other aspects of data management, new computer software is developing to streamline the presentation of reports, and the student of Statistics needs to learn to master this software.

Data roles and types

Comparative studies are designed to investigate associations between determinants and outcomes. For example in a study investigating factors affecting the duration of inter-pregnancy intervals among Malaysian women with two or more children, family income is a determinant and the interval from the time of the first to the second pregnancy is an outcome. Similarly, in a study of recidivism among heroin addicts allocated to methadone or non-methadone treatment, having a prison record is a determinant and dropping out of the program within 12 months is an outcome.

Whether a variable measured in a study is a determinant or an outcome is called its *role.* Note that a variable could play the role of a determinant for one question of interest, and the same variable could be an outcome for another question of interest. For example, in a study of risk factors for alcoholism among men, one question of interest is 'does having an alcoholic father increase the risk of being an alcoholic?' Another question is 'is alcoholism associated with early mortality?' In the first question of interest, alcoholism is an outcome, whereas alcoholism is a determinant when studying the second question.

Measurements are often classified as nominal, ordinal, or interval. *Nominal* data are categorical, like gender (male or female), marital status (single, married, divorced, widowed, etc.) and colour (blue, green, yellow, red, etc.) If there are only two possible values for a nominal measurement, it is said to be *binary*.

Ordinal data are also categorical, but have a natural ordering. Questionnaires used in opinion surveys often have ordinal responses. For example, a questionnaire could comprise a number of statements each asking for a response selected from the following choices: "strongly agree", "agree", "unsure", "disagree", "strongly disagree". These responses could be coded as 1, 2, 3, 4, 5, or perhaps in the reverse order as 5, 4, 3, 2, 1, or as –2, –1, 0, 1, 2, indicating the level of agreement.

Another example of ordinal data is a person's educational status, in terms of the amount of schooling or university studies completed,

eg. did not attend school

did not complete primary school,

completed primary school,

completed secondary school,

completed undergraduate degree,

completed postgraduate degree.

Interval data (also called *continuous* data) arise when more accurate measurements are made. For example, it may be possible to measure the amount of rainfall at a meteorological station to the nearest millimetre, or the concentration of lead in a person's blood to the nearest part per million.

Note that interval data are often rounded or grouped. For example, the age of a person is usually truncated to complete years, and sometimes grouped further into ten-year age groups, for example. So interval data may become ordinal if the number of distinct categories is small (less than 10, say).

Data may be *numerical* (that is, represented by numbers) or *categorical* (represented by names or labels). Interval data are always numerical and nominal data are always categorical. Ordinal data are usually categorical, but could be coded as integers such as 1, 2, 3,... for convenience, particularly for purposes of statistical analysis. In some situations ordinal data might be treated as continuous data to simplify the statistical analysis.

Exercise 1.2:

Classify each of the following measurements as binary, nominal, ordinal, or interval:

a person's hair colour;
the result of a pregnancy test;
the amount of income tax paid by a person;
the grade obtained by a student in a course unit;
a student's grade point average;
the material used to build a house (timber, brick, etc.);
the depth of snow at a ski resort;
the response to a questionnaire item requiring a box to be ticked;
the sex ratio of persons climbing a mountain;
the number of mosquitoes in a room.

Classification of statistical methods

Most Statistics texts cover a range of statistical methods including *z* tests for comparing proportions and *t* tests for comparing means, chi-squared tests for assessing associations, regression analysis, analysis of variance, etc. Most of these methods are covered in this book.

One of most difficult aspects faced by students learning Statistics is to choose the appropriate method from a selection of those available. But it turns out that the appropriate method may be deduced logically once the role of the variables and their data types are specified. This classification is given below in Figure 1.1.

Thus if both the determinant and the outcome variable are binary, the data may be represented as a 2×2 contingency table of counts, for which the appropriate method of analysis is the chi-squared test for independence (or, equivalently, the *z* test for comparing two proportions). Similarly, if the determinant is of the binary type and the outcome variable is continuously varying, the appropriate method is based on either a two-sample *t* test (if the data in the two determinant groups are independent samples) or a paired t test (if the samples are matched). And so it goes.

In this introductory text, we do not cover all of the methods listed in Figure 1.1. Analysis of variance methods are not covered, nor are logistic and polytomous regression. But the other methods are described in the chapters to follow. Thus two-sample and paired t tests are covered in Chapter 8, simple linear regression is treated in Chapter 9, and methods for analysing categorical data in the various contingency tables (2×2 2×c, r×2 and r×c) are given in Chapter 10.

Outcome variable type

		binary	***multi-categorical***	***continuously varying***
Determinant variable type	*binary (2- valued)*	2 × 2 tables	2 × c tables	two-sample (or paired) t tests
	multi-categorical	r × 2 tables	r × c tables	analysis of variance
	continuously varying	logistic regression	polytomous regression	simple linear regression

Figure 1.1: Classification of statistical method by role & type of variables

Summary

Statistics is part of a larger area of study, which is concerned with formulating questions of interest to society and answering these questions by doing studies. If the study is well designed, properly conducted, and informatively reported, it will provide a valid answer to a question and thus contribute effectively to society's body of knowledge. This chapter has covered the basic methods for designing and conducting studies.

Study design

The first step in the study design process involves formulating an appropriate question of interest. The question is answered by determining an unknown parameter of interest in a specified target population. The parameter could be the mean or the proportion of a single variable of interest, or a measure of the association between two variables of interest.

The next step in study design is to describe the methods to be used for selecting the subjects (individuals or, more generally, experimental units) from which the data are collected. This method of selection determines the type of study to be undertaken. If no such data are needed, the study type is said to be purely deductive. If data need to be collected, the study is said to be empirical, or inductive, because the characteristics of the population are induced from a sample of measurements. Another type of study, which combines elements of deduction and induction, is a simulation study, in which data are generated by computer simulation.

Inductive studies are said to be quantitative if they involve structured data, such as those based on questionnaire responses. In contrast, a qualitative study involves unstructured data collection, usually based on in-depth interviews of small numbers of subjects.

A quantitative study is said to be descriptive if the population parameter of interest is based on a single variable. But if the parameter of interest is the degree of association between two variables, the study is said to be comparative. Comparative studies are further classified as case-control, cohort, or cross-sectional studies according to the method of selection of subjects.

Having decided on the type of empirical study to be undertaken, the next step is to choose the sample. The sample should be unbiased, that is, it should be representative of the target population. One method of eliminating bias is to use random selection to select the sample. A sample can be made more representative by stratifying the population in subgroups (classified by demographic variables such as gender and age group), and taking random samples from these subgroups.

Measurement bias arises when an instrument for data collection gives consistently wrong results. Selection bias occurs when the sample is not representative of the target population. Response bias is a special case of selection bias that occurs when the proportion of data that are missing is unduly large: an example is individuals failing to respond to mailed questionnaires.

The accuracy with which a population characteristic is measured is determined by two factors: (a) the sample size, and (b) the extent of variation in the data in the target population. The inaccuracy is directly proportional to the variation in the data, where this variation is expressed in terms of a quantity called the standard deviation. But the accuracy does not depend on the size of the population.

Reporting results

It is important to know how to present the results of a study. The report should include effective graphs and appropriate statistical analyses. The data need to be properly stored and managed.

Data management entails effective measurement, collection, maintenance, analysis, and reporting of the data needed in a study. New technology, particularly new computer software, has made this task a lot easier.

The choice of statistical analyses can be simplified greatly once the data variables are classified according to their role (determinant or outcome) and type (categorical or continuous).

Homework problems

1. An advertisement states that three-quarters of dentists recommend Brand X chewing gum for their patients who chew gum. Comment on this claim. (How might the sample have been collected? Is bias likely?)

2. An aspiring student politician decides to assess her chances of election to the University's Student Representative Council. She decides to poll 100 (2%) of the students enrolled, by getting a list of students from the Registrar and selecting every 50th record. Her campaign manager suggests that this procedure may not give sufficient coverage of the 12 Schools making up the University, and that a more representative sample can be obtained by asking supporters to personally select and poll 2% of the students from each School. Which method is better? Explain.

3. A circular is sent to 500 trade union members, asking them to respond to the question 'Do you support or oppose uranium mining?' There are 100 responses of whom 88 'oppose', 10 'support', and two have no opinion. What conclusion can be drawn from this survey? Explain.

4. Comment on each of the following statements:

   a. An independent laboratory test showed that users of a particular brand of toothpaste had 36% less cavities.

   b. Most people have below-average incomes.

   c. Since two-thirds of all rape and murder victims are friends or relatives of their assailants, you are safer at night in a public park than you are in your own home.

   d. Since there are far more old drunks than old doctors, excessive drinking doesn't result in premature death.

   e. Statistics show that people actually get sicker, on average, after visiting a doctor. Therefore going to a doctor makes your illness worse.

5. Describe how you would go about conducting a study to determine the average annual earnings of lawyers in the city.

6. A pharmaceutical company claims it has a new cure for irritable bowel syndrome. To test its effectiveness, 50 persons are randomly selected from a group of 500 patients known to suffer from irritable bowel syndrome. After being treated for one month with the new therapy, a telephone poll reveals that 40 felt noticeably less pain, eight were unchanged, and only two patients had suffered increased pain. How would you report these results? Explain.

7. A researcher wishes to compare the promotion prospects for men and women managerial employees in large corporations. The target population for this study comprises ten corporations with 10,000 managerial employees, of whom 15% are women. The budget for the study allows for a sample size of 100, so the researcher uses a random sampling procedure to recruit 15 women and 85 men. Do you think that this allocation is reasonable? Explain.

8. To evaluate the effectiveness of a heart transplant program at a University medical centre, patients were divided into a treatment group and a control group. The treatment group comprised patients who had received a transplant, while the control group comprised patients who were recommended for a transplant but who did not receive one due to the unavailability of an appropriate donor. After two years, the survival rate in the treatment group was much higher than that in the control group. Does this result indicate that the heart transplant program was successful? Comment.

9. In New South Wales, mature-age students comprise about 45% of the undergraduate intake, and school leavers comprise about 55% (*The Sydney Morning Herald*, 23 November 1993.) From the data in Appendix B, select a sample of 50 students born in NSW and thus estimate the proportion of mature-age students. (Assume that any students from this list born before 1969 is mature-age.) Is this procedure a valid method for estimating the proportion of mature-age students at Macquarie University? Explain.

10. Suppose you wish to estimate the proportion of mature-age students at universities in NSW in the current year, and that you have a method for choosing a random sample of university students from this target population. How large a sample is needed to estimate this proportion with a probable error of 5%?

GRAPHING DATA

CHAPTER 2

2 Graphing Data

This chapter is about graphing data. The aim is to learn how to interpret data, and how to gain insight into the sources of these data. We do this by examining graphs.

Different graphs are needed for different types of data. Histograms, cumulative distribution plots, and box plots are used to display numerical data, while bar charts and pie charts are used for categorical data.

When viewing these graphs it is important to think about the data, asking questions about their source and their plausibility. Graphing data can be fun, especially if you have a good modern statistical computer package, but it is also important in business and scientific studies. One of the main reasons for graphing data is to detect discrepancies and errors in the data that could cause problems later on.

Histograms

A histogram is a way of graphically displaying a sample of numerical data. It shows how many measurements of different sizes occur (the *distribution* of the data).

In this section you will learn how to construct and interpret histograms by considering some examples, using data based on studies.

Fat percentages

Figure 2.2 shows a histogram of the following sample of body fat percentages measured on 18 adults who participated in a clinical study, reported by Mazess et al (1984). The aim of the study was to investigate a new method for measuring body composition.

9.5^1	27.9	7.8^1	17.8^1	31.4	25.9	27.4^1	25.2	31.1
34.7	42.0	29.1	32.5	30.3	33.0	33.8	41.1	34.5

Figure 2.1: Body fat measurements from 18 adults (^{1} = male)

The histogram contains rectangular bars representing the numbers of measurements in grouped intervals, referred to as *bins*. For these data, the first bin is the interval between 5 and 10: this contains two measurements (9.5 and 7.8) so the height of the bar (called the *frequency*) is 2. There are no measurements between 10 and 15, so the next bin is empty and thus its histogram bar is not shown (strictly speaking, it has frequency 0). There is a single measurement between 15 and 20 (17.8), so the next bar has frequency 1. The remaining bins have frequencies 0, 5 (corresponding to the measurements 27.9, 25.9, 27.4, 25.2 and 29.1), 8 (measurements 31.4, 31.1, 34.7, 32.5, 30.3, 33.0, 33.8 and 34.5), 0, and 2 (42.0 and 41.1).

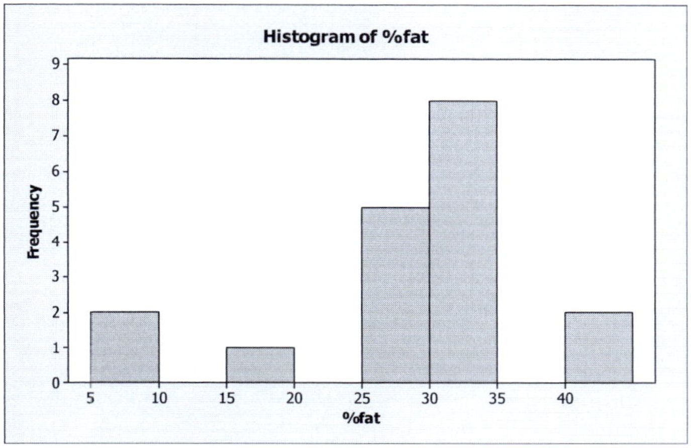

Figure 2.2: Histogram of fat percentages of 18 adults in a clinical study

When constructing a histogram you first decide on the locations of the bins, and then pass through the data list, counting the number of measurements in each bin.

If a measurement falls on the boundary between two bins, it is conventional to allocate it to the upper (or rightmost) bin. So a new subject having a fat percentage of 20.0 would have their data allocated to the bin 20-25.

The bins should start and end with 'round' numbers, like multiples of 2, 5 and 10 (possibly themselves multiplied by a power of 10, if the measurements are very small or very large).

The number of bins you choose should depend to some extent on the size of the sample. Eight bins (including those with frequencies of 0) seem to be about right for the above sample of 18 measurements. Figure 2.3 shows a histogram of the same data with the number of bins increased to 18, so that each bin covers 2 percentage points. Which do you prefer?

Of course the answer depends on what information you are trying to discover from the histogram. If you wish to see more detail you should choose narrower bins, but in doing so you may lose sight of the overall shape of the distribution.

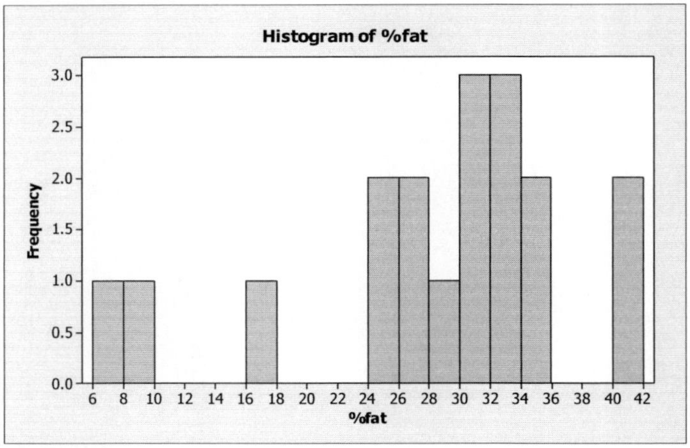

Figure 2.3: Alternative histogram of fat percentages with narrower bins

The fat percentages range from 5 to 45, approximately, with most falling between 24 and 36. The main purpose of looking at a histogram is to get information about the distribution of the data in the target population. But a sample of 18 is perhaps too small to draw such conclusions with any certainty. The next example involves a larger sample of data.

Blood lead levels

Figure 2.4 shows a histogram of blood lead levels (in micrograms/decilitre) of 433 schoolchildren living in Pattani province of Thailand in 1995. (These data arose from a study of the effects of tin mining on the environment undertaken by Alan Geater and Jiraporn Chompikul.)

In this case the bins are of width 1 unit, and some gaps and irregularities appear in the histogram. But if bins of width 2 are taken, most of these irregularities are smoothed out, as Figure 2.5 shows.

The histograms show that the blood lead concentrations range from 4 to 30 μg/dl, with the bulk of the measurements between 6 and 16 μg/dl. Since concentrations greater than 12 μg/dl are considered a health risk by the World Health Organisation, there is cause for concern: by this criterion, about half of the schoolchildren are in danger.

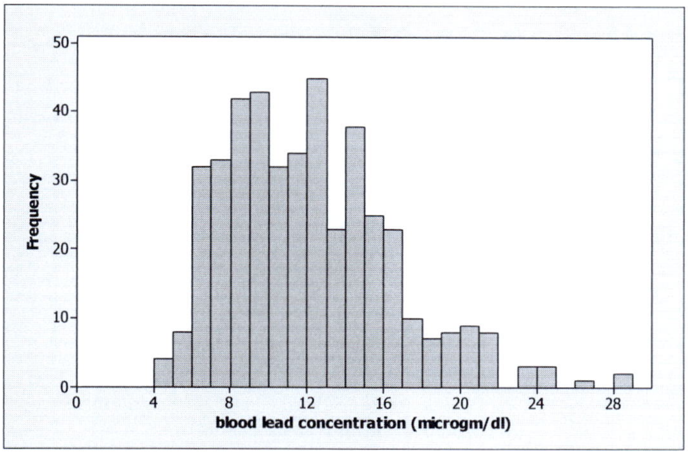

Figure 2.4: Histogram of blood lead levels of 433 schoolchildren

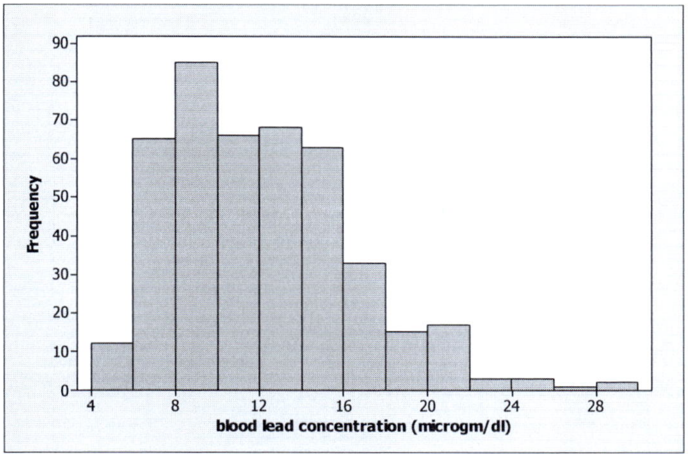

Figure 2.5: Alternative histogram of blood lead levels of 433 schoolchildren

The histogram also shows that the distribution of blood lead concentrations is skewed to the right. This result is not uncommon. Distributions of many kinds of measurements, particularly rates and concentrations, are often skewed to the right.

Stem-and-leaf plots

Histograms provide effective visual summaries of numerical data, and are used extensively in statistical data analysis. However, grouping the data into bins loses some information. If the bin width is small, less information is lost, but as a result the graph is less attractive.

A stem-and-leaf plot is similar to a histogram, but preserves the information to some extent. Consider the data listed in Figure 2.6. These are the per capita incomes (in dollars) in 1989 of the 50 states of the USA (together with the District of Columbia), as cited by Johnson (1991, p52).

State	*Income*	*State*	*Income*	*State*	*Income*
Alabama	13,679	Kentucky	13,777	N.Dakota	13,261
Alaska	21,173	Louisiana	13,041	Ohio	16,499
Arizona	15,881	Maine	16,310	Oklahoma	14,151
Arkansas	12,984	Maryland	21,020	Oregon	15,785
California	19,740	Mass.	22,196	Penn.	17,422
Colorado	17,494	Michigan	17,745	Rhode Is.	18,061
Connect.	24,604	Minn.	17,746	S.Carolina	13,616
Delaware	19,116	Miss.	11,835	S.Dakota	13,244
D.C.	23,436	Missouri	16,431	Tennessee	14,765
Florida	17,694	Montana	13,852	Texas	15,483
Georgia	16,188	Nebraska	15,360	Utah	13,027
Hawaii	18,306	Nevada	18,827	Vermont	16,399
Idaho	13,762	N.Hamps.	20,251	Virginia	18,970
Illinois	18,858	N.Jersey	23,764	Wash.	17,640
Indiana	16,005	N.Mexico	13,191	W.Virginia	12,529
Iowa	15,524	N.York	20,540	Wisconsin	16,759
Kansas	16,182	N.Carolina	15,221	Wyoming	14,135

Figure 2.6: Per capita incomes for states in the USA

As for a histogram, the first step in creating a stem-and-leaf plot is to divide the range of the data into convenient bins. Ignoring decimal places, these bins should have boundaries that are multiples of 1, or 2, or 5, and there should be about 10 to 20 such bins.

For the per capita incomes, the data range from $11,835 (for Mississippi) to $24,604 (for Connecticut), so it would be reasonable to choose bins with boundaries 11, 12, 13, ….., 24 (thousands), giving 14 bins in all.

In the stem-and-leaf plot, the bins are shown as a vertical column called the *stem*, with a vertical bar to its right, as shown in the left panel of Figure 2.7.

The second step in the construction of the stem-and-leaf plot is to allocate each data value to its appropriate stem, by writing the next significant figure to the right of the appropriate bin. For example, the first of the per capita incomes is $13,679 (for Alabama), so this value is allocated to the bin whose stem value is 13, and appears as the digit 7, because 13,679 rounds to 13,700. Similarly, the per capita income for Alaska, which rounds to 21,200, is denoted by the digit 2 placed next to the bin with stem 21. The middle panel of Figure 2.7 shows the picture after writing down the data for the first ten states in the alphabetical list (Alabama to Florida). Note that the figure for Arkansas ($12,980) rounds up to 13,000, so it is placed in the bin corresponding to 13 rather than that corresponding to 12.

After inserting each observed value in its appropriate place in the graph, the final step involves sorting the values into ascending order in each bin. This gives rise to the graph in the right panel of Figure 2.7. If the numbers in the stem have been obtained from the data by stripping a power of 10, this factor may also be included, as shown at the bottom of the graph. Minitab indicates this by specifying a leaf unit. So, for this example the leaf units would be of 100.

Step 1		*Step 2*		*Step 3*	
11		11		11	8
12		12		12	5
13		13	7 0	13	0 0 0 2 2 3 6 7 8 8 9
14		14		14	1 2 8
15		15	9	15	2 4 5 5 8 9
16		16		16	0 2 2 3 4 4 5 8
17		17	5 7	17	4 5 6 7 7 7
18		18		18	1 3 8 9
19		19	7 1	19	0 1 7
20		20		20	3 5
21		21	2	21	0 2
22		22		22	2
23		23	4	23	4 8
24		24	6	24	6
				$\times 10^3$	

Figure 2.7: Construction of a stem-and-leaf plot, for per capita incomes

Figure 2.8 shows two stem-and-leaf plots of the body weights from Figure 2.1. The graph on the left uses bins of width 5, whereas the graph in the right panel of Figure 2.8 uses bins of width 1. Note that there is some loss of accuracy in the left-hand graph, where the data displayed are rounded to integers (actually MINITAB truncates the data rather than rounding it). The graph in the right panel gives an additional digit of accuracy.

Also note, as shown in the left plot in Figure 2.8, Minitab provides a frequency column which accumulates frequencies from each end of the data up to the bin containing the median. This bin which contains the median denotes the frequency in that bin shown in brackets.

```
Leaf Unit = 1.0

 2    0   79
 2    1
 3    1   7
 3    2
 8    2   55779
(8)   3   01123344
 2    3
 2    4   12
```

```
Leaf Unit = 0.10

1   7    8
1   8
2   9    5
2   10
2   11
2   12
2   13
2   14
2   15
2   16
3   17   8
3   18
3   19
3   20
3   21
3   22
3   23
3   24
5   25   29
5   26
7   27   49
7   28
8   29   1
9   30   3
9   31   14
7   32   5
6   33   08
4   34   57
2   35
2   36
2   37
2   38
2   39
2   40
2   41   1
1   42
```

Figure 2.8: Alternative stem-and-leaf plots of body fat percentages

Exercise 2.1:

Make a stem-and-leaf plot of the per capita incomes, using bins of width 2000.

Cumulative distributions

A problem with the histogram is that the bin width needs to be defined, and as you have seen, different choices of bin width can give histograms that look rather different.

There is another way of graphing the distribution of a set of data, for which there is no need to group the data. This graph is called a *cumulative distribution.*

A cumulative distribution is a plot of the proportion of observations below x, for values of x in the range of the data.

The cumulative distribution is a step function comprising horizontal and vertical line segments. This function increases from 0 for all values less than the minimum observation, to 1 for all values greater than the maximum observation. The steps occur at the data values, and the height of each step is $1/n$, where n is the sample size.

To illustrate, look again at the body fat percentages considered in the preceding section. Figure 2.9 shows a graph of the cumulative distribution of this set of data.

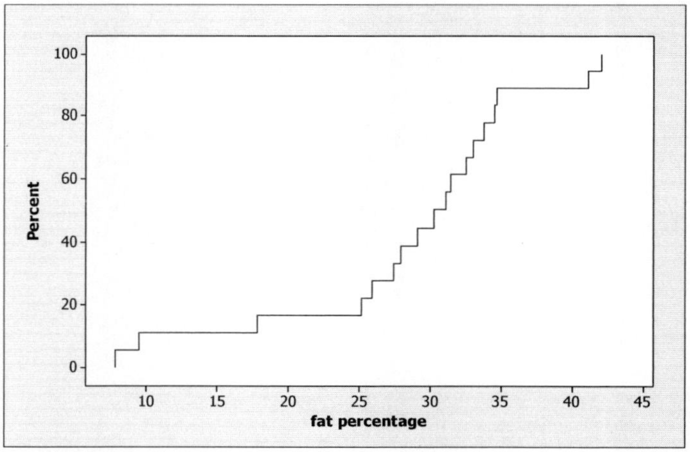

Figure 2.9: Cumulative distribution of body fat percentages of 18 adults in a clinical study

As a further illustration, Figure 2.10 shows a graph of the cumulative distribution of the blood lead concentrations of the Thai schoolchildren. In this case the curve is smoother because the sample size is larger (433).

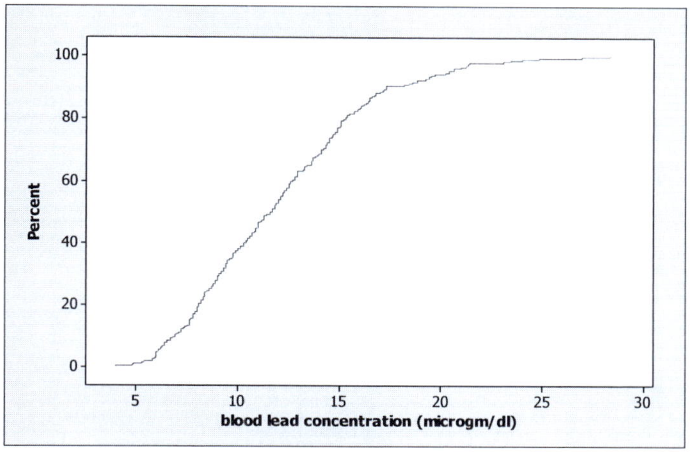

Figure 2.10: Cumulative distribution of blood lead levels of 433 schoolchildren

The cumulative distribution does not have quite the same graphical impact as a histogram in showing the distribution of the data. However, there are situations where the cumulative distribution provides useful information. In particular, the cumulative distribution facilitates the measurement of certain summary statistics of interest.

Box plots

In this section we describe another method for graphing samples of numerical data, called a *box plot*. This graph is an abbreviated plot that focuses on particular features of the distribution of the data, including summary measures called *quartiles*, defined as follows.

Quartiles

To understand how a box plot is constructed, let us reconsider the sample of fat percentages from the preceding section (listed in Table 2.1). The data, in increasing order, are as follows.

7.8	9.5	17.8	25.2	25.9	27.4	27.9	29.1	30.3
31.1	31.4	32.5	33.0	33.8	34.5	34.7	41.1	42.0

It is useful to sort the data in this way because a box plot divides the sample into four subsets each containing one quarter of the data. The values where these divisions occur are called *quartiles*. So the first quartile (referred to as the *lower quartile*) has one quarter of the measurements in the sample below it and three quarters of the measurements above it. Similarly the second quartile (the *median*) has half of the data below it and half above, and the third quartile (*the upper quartile*) has three quarters of the data below it and one quarter above.

As an exercise, cover the remainder of this page and see if you can figure out the lower quartile of the sample of 18 measurements shown above.

The answer is that the lower quartile is 25.9. Why is this?

The reason is that the sample size is 18, and 18 divided by 4 is 4½, so strictly speaking the lower quartile should have 4½ measurements below it and 18 – 4½ = 13½ measurements above it. The fifth measurement in the sorted list has 4 other measurements entirely below it and 13 entirely above it, so it satisfies the definition if you count this measurement as being half below and half above the lower quartile. So the lower quartile may be defined as the fifth measurement, that is, 25.9.

In the same way the upper quartile of the sample listed above is 33.8, the fifth largest measurement.

Now cover the rest of the page and see if you can work out the value of the median.

The answer is that the median is halfway between 30.3 and 31.1, which is 30.7. This is because the median should divide the sample into two halves. For a sample of size 18, there are nine measurements in each half, and any value between the ninth and the tenth measurements in the ordered list satisfies this definition. So it is reasonable to define the median as the average of these two middle values.

In fact, the median and quartiles are just special cases of percentiles which are defined in Chapter 3. The median is the 50th percentile and the lower and upper quartiles are the 25th and 75th percentiles respectively. The method for computing a quartile may be described as follows:

First sort the data into ascending order. Let n be the number of values in the data set. Multiply n by 25/100 (for the lower quartile since it is the 25th percentile) and let it equal m, say. If m is not an integer then the next integer after m is the position of the lower quartile. If m is an integer then take the value half-way between the mth and $(m+1)^{th}$ data values to give the lower quartile position. The median and upper quartile can be found in a similar manner.

Note that quartiles have a direct interpretation in terms of the cumulative distribution. If you draw horizontal lines on this graph corresponding to cumulative proportions 0.25, 0.5, and 0.75, the points on the x-axis where these lines intersect the curve correspond to the quartiles. This correspondence is illustrated below in Figure 2.11.

The *inter-quartile range* of a sample of measurements is simply the difference between the upper and lower quartiles. So for the above sample of 18 measurements, the inter-quartile range is 33.8 – 25.9 = 7.9.

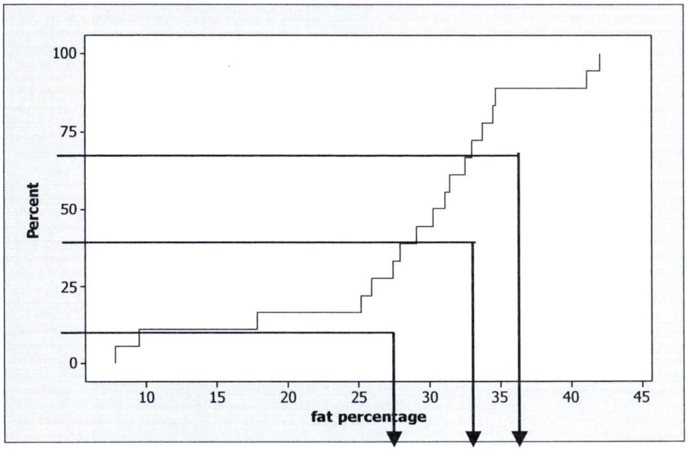

Figure 2.11: Graphical definition of quartiles of body fat percentages

Exercise 2.2:

a. *Suppose that the largest measurement in the sample of fat percentages is omitted, reducing the sample size to 17. Calculate the median and the quartiles of this reduced sample. What is the new inter-quartile range?*
b. *Suppose that the smallest two measurements in the sample of fat percentages are omitted, reducing the sample size to 16. Calculate the median, quartiles, and inter-quartile range of the reduced sample.*

Extreme values

Now we are ready to construct a box plot of the sample.

The main part of the graph is a rectangle ranging from the lower quartile to the upper quartile, and divided into two components at the median.

The remaining part of the graph comprises individually marked extreme values. These extreme data are also called *outliers*, and are defined in terms of their distance from the median, using the inter-quartile range as a yardstick. A common definition, which we will use, is to define the outliers in a box plot as measurements distant above the upper quartile, or below the lower quartile, by more than 1.5 times the inter-quartile range.

Following this definition, the outliers in the sample of 18 fat percentages are those measurements which are more than the amount 11.85 (1.5×7.9) above the upper quartile (33.8), or more than this amount below the lower quartile (25.9). This means that any outliers must be greater than 45.65 ($33.8 + 11.85$) or less than 14.05 ($25.9 - 11.85$). The only such data are the smallest two measurements in the sample (7.8 and 9.5): no measurement exceeds 45.65.

Finally, it is conventional for the box plot to have lines to be drawn from the ends of the box to the extreme measurements that are not outliers. In the present case, these data are 17.8 and 42.0.

Following this definition, Figure 2.12 shows a box plot of the sample of 18 fat percentages. This graph may be compared with the histograms of the same data shown in Figures 2.2 and 2.3. As you can see, the box plot does not focus on the shape of the distribution, but highlights any outliers that might be present.

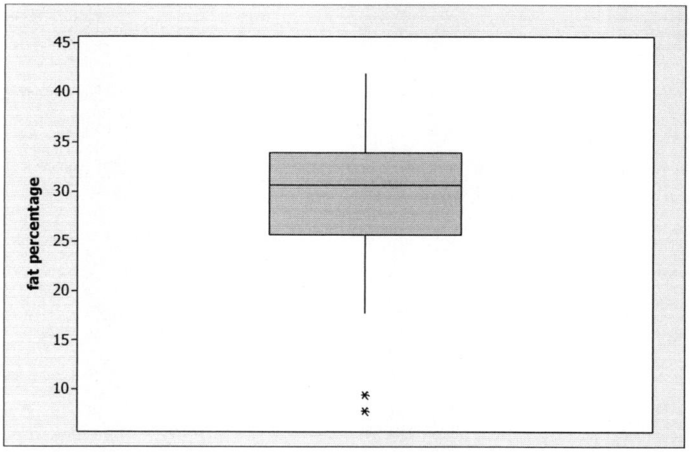

Figure 2.12: Box plot of fat percentages of 18 adults in a clinical study

Skewness and bimodality

As a further illustration, Figure 2.13 shows a box plot of the blood levels of the Thai schoolchildren considered in the preceding section. In this case there are several high outliers, and the skewness you saw in the histograms (Figures 2.4 and 2.5) is evident from the fact that the line drawn above the upper quartile is substantially longer than the line drawn below the lower quartile.

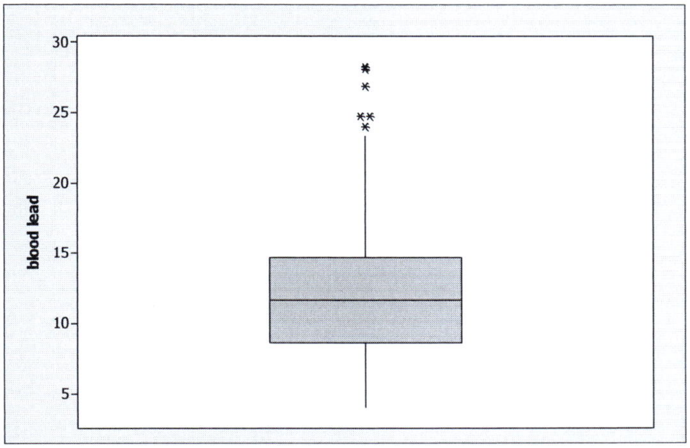

Figure 2.13: Box plot of blood lead levels (μg/dl) *of 433 Thai children*

A box plot is not always effective in showing the information in a set of data, as the next example shows. Figure 2.14 is a box plot of the direction of the current at a particular location (Hua Hin) at the north east of the Malay Peninsula during March 1994. A remote-sensing instrument recorded the measurements at various times during the day (usually at hourly intervals between 3 am and 9 am and between 3 pm and 9 pm).

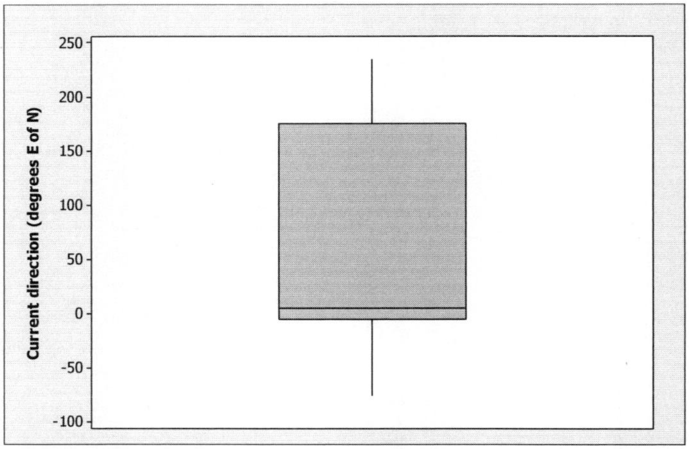

Figure 2.14: Current direction (degrees E of N) at Hua Hin in March 1994

Based on the box plot, you may conclude that the median current direction is slightly greater than 0 (north) and that half of the time its direction is between –5 and 170 degrees. You might also conclude that this distribution is skewed, because the median is displaced to the bottom of the box. (If a distribution is exactly symmetric, the median is halfway between the lower and upper quartiles.)

However, a histogram of the data gives a rather different picture, as can be seen from Figure 2.15. This graph shows that the distribution of the current direction has two components in opposite directions: one component is clustered around 5 degrees west of north and the other component is clustered around 175 degrees east of north (that is 5 degrees east of south).

A distribution with two identifiable clusters is said to be *bimodal.*

There is a simple scientific explanation for the bimodality of the distribution of the current direction. The direction of the current in large bodies of water on the Earth's surface is determined largely by the tide (which in turn is largely determined by the Moon). As a consequence of the laws of physics, when the tide is receding the water tends to flow in the opposite direction to its direction when the tide is rising.

Two conclusions may be drawn from this example.

First, box plots are not as informative as histograms for displaying the frequency distribution of a sample of data. Second, it helps a lot if you have some subject matter knowledge when interpreting graphs of data.

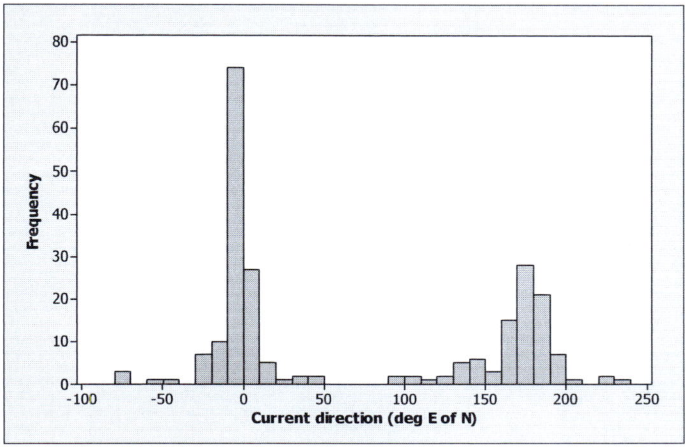

Figure 2.15: Histogram of current direction at Hua Hin in March 1994

Multiple plots

Box plots are useful for showing the distributions of several samples of similar measurements, because they are more concise than histograms.

Consider again the concentrations of lead in the blood samples taken from the schoolchildren in southern Thailand. A major objective of this study was to see if children living closer to a river contaminated by mining activity are at greater risk than those living at a greater distance from the river. The blood samples were taken from children attending five schools in villages at different distances from the river.

Figure 2.16 shows box plots of the samples from each of the five village schools. From this graph it appears that the children in schools 2, 3 and 4, which happen to be in villages closest to the river, have higher levels of lead in their blood than the children do at the other schools, on average. This provides evidence in support of the hypothesis that the children living closer to the river are at greater risk than those living further away.

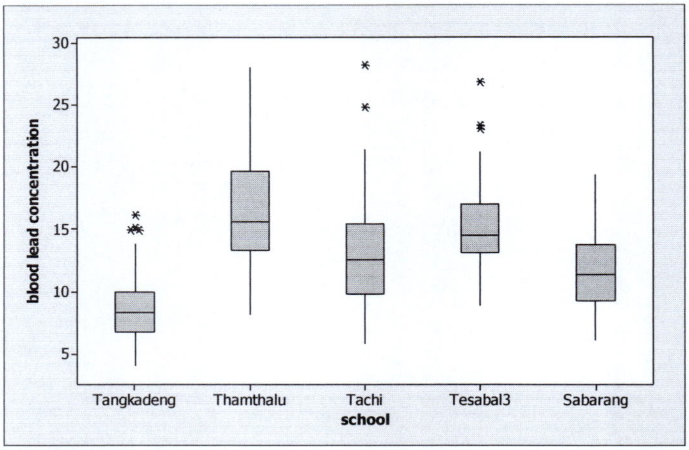

Figure 2.16: Multiple box plots of children's blood lead levels in micrograms per decilitre for each of five schools

As a further illustration consider again the body fat percentages of the persons graphed in Figure 2.2. These subjects include four males (those with responses 7.8, 9.5, 17.8 and 27.4) and 14 females, and Figure 2.17 displays separate box plots for the two sexes.

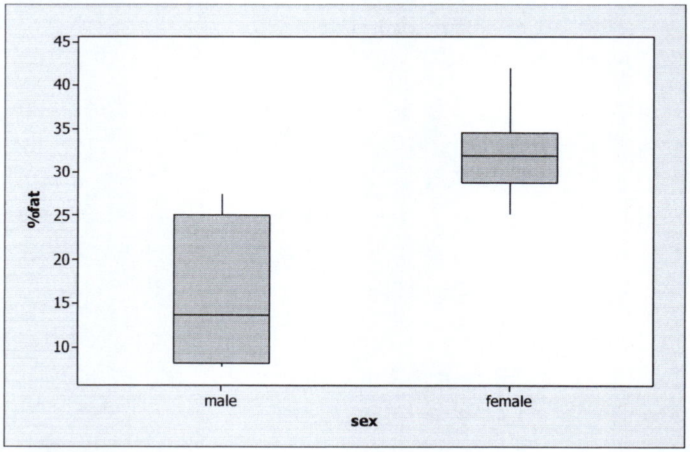

Figure 2.17: Box plots of fat percentages by sex

It seems clear from Figure 2.17 that males tend to have lower body fat percentages than females. In the following chapters you will learn how to make a more definitive comparison of the means of two groups, using useful statistical concepts called 'p-values' and 'confidence intervals'.

Bar charts and pie charts

Mei Wang did a study of knee injuries suffered by skiers as part of her Master of Applied Statistics at Macquarie University. A doctor in the medical centre at Perisher Valley classified the nature of each injury soon after it occurred. Of 492 knee injuries, 171 involved the medial cruciate ligament (MCL) but not the anterior cruciate ligament (ACL), 62 involved the ACL but not the MCL, 66 involved both the MCL and the ACL, 90 involved other parts of the knee and 96 did not involve physical damage.

These data are nominal and not ordinal. A *bar chart* could be used to graph the frequencies, as in Figure 2.18.

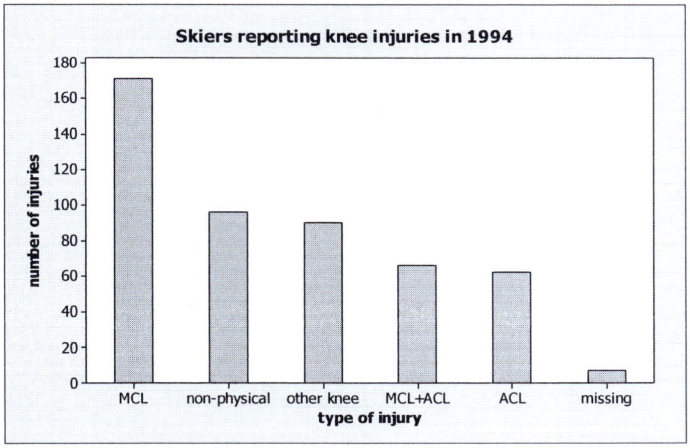

Figure 2.18: Bar Chart of Knee Injuries of Skiers

As you can see, a bar chart is similar to a histogram in that it represents the frequencies of the data as bars. The main difference is that the bars in a bar chart are separated and given text labels, whereas the bars in a histogram cover a continuous interval with numerical labels. Another difference is that histograms usually have vertical bars while bar charts can have horizontal bars (giving more room for the labels). However, it is not necessary to follow this convention and MINITAB does not offer this option of horizontal bars.

Since nominal data such as these do not necessarily have a natural order, you need to decide how to order the categories in a bar chart. You could choose to order the categories alphabetically by the categorical label, but there may be a more appropriate ordering. Ordering by frequency is common and is useful for comparison, particularly with a large number of categories.

Categorical data may also be displayed using a pie chart, as shown in Figure 2.19. In this graph the counts in the various categories are represented as sectors of a circle, in which the area of each sector is directly proportional to its percentage, and the sectors are ordered in the clockwise direction starting from the top.

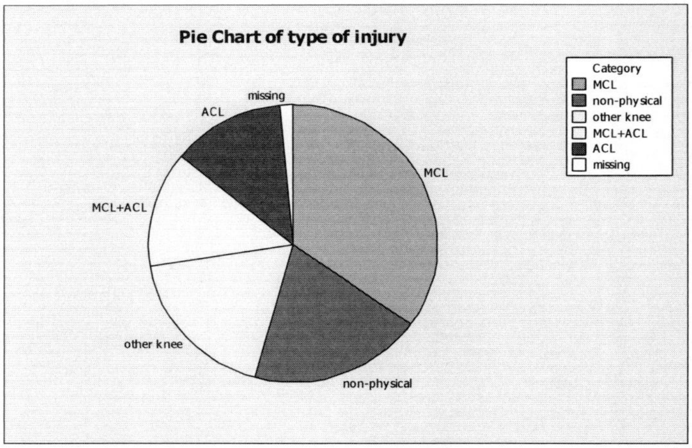

Figure 2.19: Pie Chart of Knee Injuries of Skiers

A pie chart may be preferable to a bar chart if the percentages in the various categories are of primary interest. However, a bar chart is preferable for comparing the counts or percentages in the various categories with each other. Thus, for example, the pie chart of the skiing injuries clearly indicates that medial cruciate ligament injuries make up approximately one third of all knee injuries, whereas it is difficult to see this fact from the bar chart.

On the other hand, the bar chart shows clearly that injuries involving both the medial and anterior cruciate ligaments are a little more frequent than injuries involving only the anterior cruciate ligament. This is less obvious from the pie chart.

Summary

In this chapter you have seen how data may be graphed to show their distribution. For numerical data, histograms, stem-and-leaf plots, cumulative distribution plots, and box plots are appropriate.

A histogram is a graph comprising vertical bars showing the frequency of the data in intervals of fixed width (called bins) over their range. If the data comprise a sample from some target population, the histogram gives an indication of how the data in this target population are distributed. It is thus important to choose histogram bins that are not too wide or too narrow, and thus give an effective summary of the distribution.

A stem-and-leaf plot is similar in appearance to a histogram, but gives the data individual labels corresponding to their digits. It is designed primarily for making graphs by hand, and has been largely superseded with the availability of computer technology.

Some information is usually lost in a histogram or stem-and-leaf plot, because to produce these graphs the data must be rounded to two or three significant figures. A cumulative distribution plot shows the exact proportion of data less than x for all values of x in the range of the data. It is thus a step function that increases by small amounts at points corresponding to the data. However, a cumulative distribution plot is less effective than a histogram in highlighting the distribution of the data.

Box plots are graphs that summarise the data. They show the medians, quartiles, and extreme values. They are very useful for comparing two or more sets of numerical data.

Bar charts and pie charts are used to display categorical data.

Homework Problems

1. The sulphur dioxide content in the air (in micrograms per cubic metre) is measured at environmental monitoring stations at 41 cities in the U.S. The following data, cited by Sokal & Rohlf (1981, page 239), are the averages taken over the years 1969-71. Make a stem-and-leaf plot of these data, and comment on its shape.

10	13	12	17	56	36	29	14	10	24	110	28	17	8	30	9
47	35	29	14	56	14	11	46	11	23	65	26	69	61	94	10
18	9	10	28	31	26	29	31	16							

2. Higham, Kijngam & Manly (1980, page 4) gave the mandible lengths in millimetres for 20 golden jackals (*Canis aureus*) from the collection of the British Museum (Natural History) as follows, with ^{1} indicating males. Make a histogram of these data.

   120^1 107^1 110^1 116^1 114^1 111^1 113^1 117^1 114^1 112^1
   110 111 107 108 110 105 107 106 111 111

3. For the data in Problem 2, draw separate box plots for the two sexes.

4. Sokal and Rohlf (1981, page 239) reported the following per diem fecundities (average numbers of eggs laid per female per day for the first 14 days of life) for 25 females of each of three genetic lines of the fruitfly *Drosophila melanogaster*. Make box plots of these data and comment on the resulting graph.

   RS: 12.8 21.6 14.8 23.1 34.6 19.7 22.6 29.6 16.4 20.3 29.3 14.9 27.3 22.4 27.5 20.3 38.7 26.4 23.7 26.1 29.5 38.6 44.4 23.2 23.6

   SS: 38.4 32.9 48.5 20.9 11.6 22.3 30.2 33.4 26.7 39.0 12.8 14.6 12.2 23.1 29.4 16.0 20.1 23.3 22.9 22.5 15.1 31.0 16.9 16.1 10.8

   NS: 35.4 27.4 19.3 41.8 20.3 37.6 36.9 37.3 28.2 23.4 33.7 29.2 41.7 22.6 40.4 34.4 30.4 14.9 51.8 33.8 37.9 29.5 42.4 36.6 47.4

5. Pocock (1983, page 191) reported the results from a clinical trial comparing two treatments for patients with lymphomas, as follows. Of 273 patients, 57 were classified as having a 'complete response', 110 as having a 'partial response', 32 'no change', and 74 'disease progression'. Make a bar chart of these data.

6. Hand et al (1994, page 237) cited Pearson's 1909 classic data on crime and drinking as follows. These are the numbers of persons convicted of the six crimes listed. Display these data using two pie charts.

Crime	*Drinker*	*Abstainer*
Arson	50	43
Rape	88	62
Violence	155	110
Stealing	379	300
Coining	18	14
Fraud	63	144

7. The results of a survey of 426 students at Macquarie University taking a first-year Statistics unit in 1988 are listed in Appendix B.
   a. Select a random sample of 30 students and record their drink preferences. Construct
      i. a bar chart and
      ii. a pie chart of the data in this sample,

      and comment on their appearances.
   b. Record the number of hours spent viewing TV by the persons with IDs 81 to 100. Graph these data using a histogram.
   c. Construct a stem-and-leaf plot of the weights of every tenth woman in the list, and describe the graph. Is this sampling procedure valid?

8. Dale et al (1987) analysed the following beta-endorphin concentrations (in pmol/litre) of runners who participated in a half-marathon 'fun run' in Britain. (Data supplied by Dr Mike Campbell, University of Southampton.) The before-after measurements come from eleven runners who completed the race, while the other column comprises the measurements taken after their participation from eleven different runners who collapsed during the race.

Before run	*After run*	*Collapsed*
6.6	24.0	177.0
4.4	30.0	160.0
8.4	20.2	71.2
7.2	37.0	66.0
17.4	46.2	414.0
5.2	15.6	107.6
9.0	20.9	83.0
4.6	25.0	122.0
6.6	24.0	177.0
4.4	30.0	160.0
8.4	20.2	71.2

   a. Make box plots of these data, using a common horizontal scale.
   b. Repeat the graph after taking logarithms of the data, and comment on these graphs. Which graph is more appropriate?

9. In an advertisement on page 52 of *The Australian* newspaper on 11 November 1997, the following salaries (in thousands of Australian dollars) for computer professionals in Sydney were given. Make a stem-and-leaf plot of these data.

   85, 70, 50, 35, 42, 30, 70, 77, 67, 42, 45, 45, 45, 45.

10. According to the 1985 Science Citation Index, the following members of British University Psychology departments had 60 or more journal citations. (The numbers of citations are in parentheses.) Make an appropriate graph of these data, and comment.

J.M. Argyle (170)
D.A. Booth (101)
O.J. Braddick (83)
T.G.R. Bower (85)
M. Coltheart (164)
S.J. Cooper (82)
H.J. Eysenck (813)
M.W. Eysenck (90)
S.B.G. Eysenck (91)
A.F. Furnham (62)
H. Giles (70)
J.A. Gray (251)
D.N. Lee (84)
N.J. Mackintosh (176)
T.W. Robbins (69)
E.T. Rolls (61)
J. Sandler (97)
H.R. Schaffer (67)
C.B. Trevarthan (73)
M.R. Trimble (97)
P.H. Venables (68)
P.B. Warr (120)
E.K. Warrington (180)
L. Weiskranz (64)
G.D. Wilson (84)
W. Yule (60)

SUMMARISING DATA

CHAPTER 3

3 Summarising Data

In the preceding chapter we looked at some ways of graphing data, with the emphasis on showing the whole distribution of the data. For numerical data you saw how histograms and stem-and-leaf plots are effective graphs but need the bin width to be defined appropriately, and you saw how cumulative distributions avoid the bin width definition problem but have less visual impact. Bar charts and pie charts are effective for displaying nominal data classified into categories.

Box plots are useful because they are more concise than histograms and cumulative distributions, and thus can show the distributions of several samples at the same time. But they don't show all the data: they summarise the data in each sample with just a few summary measures, comprising the median, quartiles, and outlying extreme values.

In this chapter we focus on data summarisation. This is necessary because there is usually more data collected in a study than can be easily digested at one time, so it is important to be able to focus on the salient features containing the relevant information.

When summarising numerical data, measures of spread are needed as well as definitions of typical or central values. For nominal measurements, the proportions in specified categories of interest are the statistics of most importance. Percentiles also play an important role when summarising numerical data. Methods for calculating these statistics are covered in the present chapter.

Typical values

In this section we talk about using a value to describe the *typical* measurement in a sample. The idea is to summarise the data using just one number. The number chosen should be *central*, that is, it should be close to the middle of the distribution. There are two obvious choices, the *median* and the *mean*. These are useful statistics for summarising numerical data measured on an interval or continuous scale, but are not appropriate for categorical data (such as ordinal or nominal data).

For categorical data the *mode* is sometimes used as a single summary statistic. The mode is simply the measurement occurring most frequently in the sample. However, the proportion of observations within a specified category of interest (which may or may not correspond to the mode) is a more useful way of summarising categorical data.

The median

In Chapter 2, when constructing a box plot, you saw how to define the median. The median is the middle value of a set of ordered numbers.

Suppose a nurse took three successive measurements of your diastolic blood pressure (in mm) and got the values 78, 81 and 80. The median is thus 80. If the sample has an even number of measurements, the median is taken as the average of the two in the middle. So if a fourth measurement of your blood pressure gives the reading 92, the median of the four measurements is 80.5 (the average of 80 and 81, the middle two measurements). The median provides a typical value for a sample of data.

It is useful to have a typical value to summarise the data in a sample, because this gives a very concise summary. Just as a box plot of a sample provides a more concise graph than a histogram, the median provides an even briefer description. You cannot get anything more concise than a single typical value.

Consider some of the examples from the preceding chapter.

The median of the body fat percentages is 30.7. So you could say that a typical percentage of body fat in an adult is 30.7. This estimate is based on a sample of 18 persons participating in a clinical study.

The median of the schoolchildren's blood lead concentrations is close to 12 (see Figure 2.13). So you could say that a typical level of lead in the blood of a schoolchild in Southern Thailand is 12 micrograms per decilitre. This estimate is based on a sample of 433 children selected from those living in five villages.

The median direction of the current at Hua Hin is 5 degrees east of north (see Figure 2.14). This estimate is based on a sample taken at specified hours during the day in March 1994.

The median is an estimate of the *centre* of a distribution. It contains no other information about the distribution. The amount of variation in the data may be relatively small (as for an individual's blood pressure, which usually does not vary by more than one or two percent) or large (as for the current direction, which embraces a range of almost 360 degrees).

The mean

Another measure of the centre of a distribution is the arithmetic *mean*, or average. The mean is quite easy to compute: you just add up all the measurements and divide by the sample size. For example, the mean of the three hypothetical blood pressure measurements (78, 81 and 80) is (78+81+80)/3, that is, 239/3 = 79.67. This is quite close to the median (80).

The mean and the median are not necessarily close to each other. If the fourth blood pressure reading (92) is included, the sample mean is now (78+81+80+92)/4, that is, 331/4 = 82.75. This is no longer close to the median (80.5). Including the additional measurement has increased the mean by more than 3 units, but has only increased the median by 0.5.

Which typical value is better to use, the median or the mean? The answer is that it depends very much on the distribution of the data. You can see from the example involving the blood pressure measurements that the median is less affected by atypical values in the sample.

The fourth blood pressure reading is more than 10 units greater than the others. The first three readings are all within 3 units of each other, ranging from 78 to 81. If these are measured correctly, it seems unlikely that another measurement taken a short time later will be so much higher than the previous one. It could be an error. It is certainly unusual.

So the median is a better estimate of the typical value in a sample if you do not wish to be swayed by extreme measurements. But the mean gives the same weight to every measurement in the sample, so it can be influenced unduly by extreme measurements.

To emphasise the difference between the mean and the median, suppose that the fourth blood pressure measurement was recorded as 820 instead of 92. Of course it is impossible for a person to have such a high blood pressure, but just suppose that a data entry clerk erroneously recorded this value. Then the median is still 80.5 (the average of the middle two, 80 and 81), but the mean has skyrocketed to (78+81+80+820)/4 = 264.75. So the median can cope with the error, but the mean cannot.

Does this mean that the median is preferable to the mean for summarising a sample of numerical data? Not necessarily. The mean is often quicker to compute than the median, because it does not require the data to be ordered. And provided the data are free of errors, statistical theory shows that the mean is a more accurate measure of the central value in a population. For this reason, investigators doing quantitative studies usually prefer the mean.

The median is easy to understand: it is simply the value that divides the sample into two equal halves, so that half the data are below the median and half are above it. Does the mean have a simple interpretation too?

If you've studied physics or surfed or roller-skated, you know about balance. Skiers maintain their balance by keeping their centre of gravity between their skis, just as a waitress in a beer hall can balance a tray of drinks by holding it under its centre of gravity.

The mean of a distribution is located at the centre of gravity of its histogram. Figure 3.1 shows the histogram of the sample of 18 body fat percentages seen earlier as Figure 2.2, together with the sample mean indicated by the arrow. The mean of this sample of measurements is

$$\frac{9.5 + 27.9 + 7.8 + 17.8 + 31.4 + 25.9 + 27.4 + 25.2 + 31.1 + 34.7 + 42.0 + 29.1 + 32.5 + 30.3 + 33.0 + 33.8 + 41.1 + 34.5}{18}$$

This is 515.0/18 = 28.61.

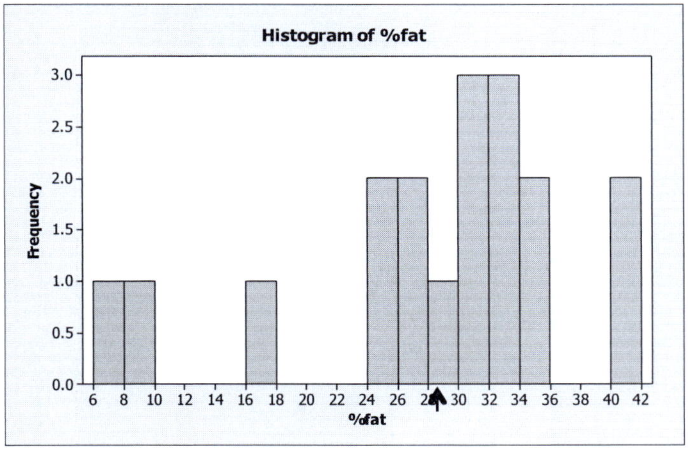

Figure 3.1: Histogram of 18 fat percentages, with sample mean indicated

Continuing with the physics analogy, the median is the point in the histogram having equal mass on each side. (Note that this is not the same as having equal balance on each side, as you can see by imagining three persons balancing on a seesaw.) The median of the fat percentages is 30.7, somewhat greater than the mean.

The more symmetric a distribution is, the closer together are its mean and median. A sample having a perfectly symmetric distribution has the same mean and median. If the distribution is skewed to the left its mean is located to the left of its median (like the data graphed in Figure 3.1). Conversely if a distribution is skewed to the right its mean is located to the right of its median. The blood lead concentrations of the Thai schoolchildren (see Figures 2.4 and 2.5) have a right-skewed distribution. The mean of this sample is 12.05, which is slightly greater than the median, 11.7.

The symbol $\bar{y}$ denotes the mean of n measurements (y_1 , y_2, …, y_n). That is,

$$\bar{y} = \frac{1}{n}\left(y_1 + y_2 + \ldots + y_n\right) = \frac{1}{n}\sum_{i=1}^{n} y_i \qquad (3.1)$$

Proportions and percentages

If the data are not numerical but categorical it is more difficult to define a typical value. Demographers sometimes try to summarise the results of a survey by describing the typical respondent, using the mean or median to describe continuously varying response data and the mode for nominal data. This might give the answer 'a male aged 33, married, with 1.8 children, earning $37,500 per annum' as the typical respondent. While this answer may be satisfactory for the numerical variables age, number of children, and income, it is not very informative to say that the typical respondent is a married male, because this gives little indication of the percentage of males or married people in the sample.

When summarising categorical data it is better to choose a category of interest and to give the proportion of observations in this category.

When the data are binary (such as gender) this approach is fully informative. Thus saying that the proportion of males in a study is 60% gives a lot more information than saying that the typical respondent is male. If another survey gave 90% males, you would be able to say that there is a much greater proportion of males in the second survey, but you would not be able to reach such a conclusion using the 'typical response' method.

For nominal data with more than two categories, you may choose as many response categories as you require, and then give the proportions of observations in these categories. For brevity you may choose to give the proportion in just one category, such as the number of persons 'currently married' in a survey of immigrants. Or you may choose two categories of interest, such as 'cancer' and 'heart disease' among causes of death.

When summarising categorical data, a major advantage of a proportion is that it is numerical. So a proportion is a method for transforming categorical data to numerical data.

There is no essential difference between a proportion and a percentage. A percentage is expressed as the number of observed cases per 100, whereas a proportion is the number of observed cases per unit. In demographic and epidemiological applications, vital rates are often expressed as the number of observed cases per thousand.

Percentiles

In Chapter 2, the method given for constructing a box plot uses the median and the lower and upper quartiles. These values divide a set of data into quarters with the lower quartile cutting off the bottom quarter, the median cutting the data in half and the upper quartile cutting off the top quarter. These are respectively the 25th, 50th and 75th percentiles. Given these definitions, it should be possible to find any percentile of a set of data.

The Pth percentile of a set of data is simply the value that exceeds P percent of the data in the set. In other words, if the Pth percentile is x, then P percent of the data values are below x and $(100-P)$ percent of the values exceed x.

For example, consider the way in which high school students in the State of New South Wales used to receive their final examination mark at the end of their schooling. Most students were supplied with a Tertiary Entrance Rank (TER), as well as an outline of marks for each individual subject for which they had been examined. This TER gave each student a ranking which enabled comparison with all other students in NSW who opted to take a TER. For example, a student with a TER of 65.15 scored better than 65.15% of other students in NSW. This system was replaced by a different score, namely the UAI (Universities Admission Index) and has now been replaced by an ATAR (Australian Tertiary Admission Rank) score.

In the preceding chapter a method was given for computing quartiles. For example, the lower quartile is defined as a value (x, say) dividing an ordered set of data into two components, with one quarter of the data below x and the remainder above x. You also saw (in Figure 2.11) a graphical technique for computing the quartiles of a set of data.

The method for computing the Pth percentile is similar. It is a value x that divides an ordered set of data into two components so that P percent is below x and the remainder is above x. Similarly, a percentile is computed graphically from a plot of the cumulative distribution, by drawing a horizontal line from the point corresponding to the proportion $P/100$ on the y-axis to the curve, and then dropping a vertical line to the x-axis.

Figure 3.2 gives an illustration of this graphical procedure, showing the 90th percentile of the blood lead concentrations of the Thai schoolchildren, which is 17.3.

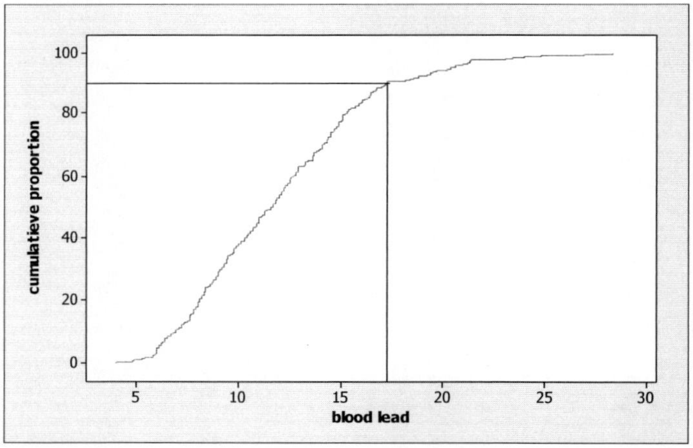

Figure 3.2: 90th percentile of blood lead concentrations of 433 schoolchildren

The method for computing medians and quartiles was described in Chapter 2. The method for computing any percentile may be described as follows:

First sort the data into ascending order. Let n be the number of values in the data set and P the required percentile. Multiply n by $P/100$ and let m be the nearest integer greater than or equal to the resulting value. If $n \times P/100$ is not an integer, the required percentile is the mth observation in the sample starting from the smallest. If $n \times P/100$ is an integer, the required percentile is halfway between the mth and $m+1^{\text{th}}$ observation in the ordered sample.

So to get the 90th percentile of the sample of lead concentrations, first note that P is 90 and n is 433. So multiply 433 (n) by 0.9 ($P/100$) to get 389.7. The next whole number is 390 (m). The 390th observation in the ordered sample is 17.3. This is the 90th percentile.

As a further illustration, consider again the ordered sample of body fat percentages, and suppose you need to compute the 33 ⅓ percentile.

7.8	9.5	17.8	25.2	25.9	27.4	27.9	29.1	30.3
31.1	31.4	32.5	33.0	33.8	34.5	34.7	41.1	42.0

Note that the sample size n is 18 and P is 33 ⅓, so $n \times P/100$ is $18 \times \frac{1}{3} = 6$, m is a whole number and consequently the percentile is halfway between the 6th and 7th data values in the ordered list. The 6th data value in the ordered list is 27.4 and the 7th is 27.9, so the 33-and-a-third percentile is their average, 27.65.

If P is close to 100 it is easier to compute the percentile by starting with the largest observation and counting down. This involves multiplying n by $(100-P)/100$, again taking the nearest greater integer m, and counting the mth observation from the top of the ordered list. To get the 90th percentile of the fat percentages, P is 90, so multiply 18 by 0.1 to get 1.8 giving $m = 2$, and note that the second largest observation is 41.1.

Exercise 3.1:

a. Use the first method given to check that the 90th percentile of the fat percentages is 41.1

b. Suppose that two more persons with body fat percentages of 30.5 and 42.5 are added to the sample. Now calculate the 20th and 80th percentiles.

Exercise 3.2:

What is the 0th percentile of a set of data? And what is the 100th percentile?

Measures of spread

We began this chapter by defining two typical values to describe a set of data, namely the mean and the median. But is this enough? Consider the following two sets of data representing the ages of two groups of students at Macquarie University who took an introductory statistics unit:

Day Students:	17	18	19	19	25	31	31	32	33	45
Evening Students:	24	24	25	26	28	28	28	29	29	29

It turns out that the means of the two samples are both 27 and the medians are both 28. But you can see that the two samples are rather different. What is different about them? Clearly the ages of the students taking the day unit are much more spread out than the ages of those taking the evening unit.

So, as well as a measure of centre, we need another measure which says something about the variability in the data. In fact, there are three different commonly used measures of spread for numerical data: the *range,* the *midspread,* and the *standard deviation.*

The range

The range of a given set of data is simply the difference between the largest and the smallest observation.

Consider some of the examples outlined in Chapter 2. The range of the body fat percentages is the difference between the highest observed value (42.0) and the lowest observation (7.8), that is, 42.0 −7.8 = 34.2. The range of the children's blood lead concentrations, shown in Figure 2.4, is 28 – 4 = 24 micrograms/decilitre.

The range is easy to understand, but is not always an effective measure of the spread in a sample because it is totally dependent on the extreme observations. Consider the ages of the day students listed above. The range is 45−17 = 28 years. However, if the 45 year-old student is omitted from the sample, the range is reduced to 33 – 17 = 16 years.

Consequently, it would be useful to have a measure of spread that is not unduly influenced by extreme values. Such a measure is described in the next section.

The midspread

The *midspread* or *inter-quartile range* is an alternative measure of spread that is not influenced by extreme values. As its name suggests, the midspread describes the middle 50% of the data, omitting 25% at each end.

Since the quartiles cut off the bottom and top 25% of the data, the difference between the two quartiles defines the midspread. For the day students' ages, the quartiles are 19 and 32, so the midspread is 32 – 19 = 13 years, clearly indicating the difference in spread compared with the evening students' age midspread of 29 – 25 = 4 years. Note, however, that the midspread of the day students, after removing the extreme value, is 31–19 = 12 years, highlighting the fact that the midspread is not influenced by the extreme value of 45 years.

Looking back at the data from Chapter 2, the midspread of the body fat percentages is 33.8–25.9 = 7.9, and, reading from the box plot graphed in Figure 2.13, the midspread of the children's blood levels is approximately 15 – 8 = 7 μgm/dl.

The standard deviation

Another measure of variation is the *standard deviation*. The mean and the standard deviation are most commonly used to summarise a set of data.

The *standard deviation* can be defined as an average deviation from the mean. Consider subtracting each of the data values from their mean. This can be done without affecting the spread of the data. The new values obtained which are referred to as the *deviations from the mean* or, more simply, the *residuals,* are centered at zero and thus have a mean of zero.

In order to calculate the standard deviation of a set of data we shall work with the *residuals*. The formula for calculating this standard deviation is

$$s = \sqrt{\frac{S_{YY}}{n-1}}, \tag{3.2}$$

where S_{YY} is the sum of the squares of the residuals, that is,

$$S_{YY} = (y_1 - \bar{y})^2 + (y_2 - \bar{y})^2 + \ldots\ldots + (y_n - \bar{y})^2 = \sum_{i=1}^{n}(y_i - \bar{y})^2 \tag{3.3}$$

Note that each residual is squared, otherwise the sum adds to zero. So it is necessary to take the square root at the end to balance things out. Also note that we have divided the number of squared residuals by $n-1$ rather than by n. This is because we know that the sum of the residuals is zero, so once we know $n-1$ of them we can calculate the nth residual by subtraction. We therefore have only $n-1$ *independent* terms in the sum.

We shall use the formula described above to calculate the standard deviation of the ages of the day students, as follows:

	Data Value y	Residual $y \ \bar{y}$	Squared Residual $(y \ \bar{y})^2$
	17	$17 - 27 = -10$	$(-10)^2 = 100$
	18	$18 - 27 = -9$	$(-9)^2 = 81$
	19	$19 - 27 = -8$	$(-8)^2 = 64$
	19	$19 - 27 = -8$	$(-8)^2 = 64$
	25	$25 - 27 = -2$	$(-2)^2 = 4$
	31	$31 - 27 = 4$	$(4)^2 = 16$
	31	$31 - 27 = 4$	$(4)^2 = 16$
	32	$32 - 27 = 5$	$(5)^2 = 25$
	33	$33 - 27 = 6$	$(6)^2 = 36$
	45	$45 - 27 = 18$	$(18)^2 = 324$
Sum	270	0	730

Using the formula (Equation 3.2) the standard deviation is

$$s = \sqrt{\frac{S_{YY}}{n-1}} = \sqrt{\frac{730}{9}} = 9.01$$

An advantage of the standard deviation is that it can be easily found using a calculator. More importantly, and as you will see in the chapters to follow, it is used, together with the sample mean, for making inferences about the population mean. Its main disadvantage, however, is that it is strongly influenced by outliers.

Consider again the sample of ages of the day students, but this time omitting the 45 year-old student. The recalculation of the standard deviation is indicated in the following table.

	Data Value y	Residual $y\ \bar{y}$	Squared Residual $(y\ \bar{y})^2$
	17	$17 - 25 = -8$	$(-8)^2 = 64$
	18	$18 - 25 = -7$	$(-7)^2 = 49$
	19	$19 - 25 = -6$	$(-6)^2 = 36$
	19	$19 - 25 = -6$	$(-6)^2 = 36$
	25	$25 - 25 = 0$	$(0)^2 = 0$
	31	$31 - 25 = 6$	$(6)^2 = 36$
	31	$31 - 25 = 6$	$(6)^2 = 36$
	32	$32 - 25 = 7$	$(7)^2 = 49$
	33	$33 - 25 = 8$	$(8)^2 = 64$
Sum	225	0	370

The standard deviation of the truncated sample is thus

$$s = \sqrt{\frac{S_{YY}}{n-1}} = \sqrt{\frac{370}{8}} = 6.80$$

Omitting a single outlier thus reduces the standard deviation from 9.01 to 6.80. This is greater than the effect of the outlier on the mean, which is reduced from 27 to 25.

Exercise 3.3:

Calculate the range, midspread and standard deviation of the ages of the evening students.

Quick rules of thumb

It is useful to be able to quickly summarise data, without going to the trouble of getting exact results. In this section we give some useful rules for handling numerical data.

Finding the median and quartiles from a stem-and-leaf plot

If the data are already displayed in a frequency graph such as a stem-and-leaf plot, you can get the median and the quartiles quickly by counting the digits shown in the graph.

For example, for the data shown in Figure 2.6, we can obtain the lower quartile approximately from Figure 2.7. In this case the sample size is 51, so the lower quartile is the 13th value (51/4 = 12.75, rounded up to the next whole number) in the ordered list. This value is the last digit on the third line of the stem-and-leaf plot shown in the right panel of Figure 2.7, which is 13,900. The exact value is that for Montana (13,852).

Estimating the mean and the standard deviation

If the data in the sample are approximately symmetric, a very rough estimate of the mean or median is the value in the middle of the range. This is the same as the average of the minimum and maximum values. For example, the body fat percentages listed in Figure 2.1 have minimum and maximum values 7.8 and 42.0, respectively, giving an estimate (7.8+42.0)/2 = 24.9 for the mean and the median. The exact values are 28.61 for the mean and 30.7 for the median, so the rough estimates are in the right ballpark.

Of course if the data have a skewed distribution this method doesn't work very well and in any case the mean and the median are different for skewed data. Outliers can also wreck the accuracy of such calculations.

You can also get a quick rough estimate of the standard deviation of a sample. You simply divide the range of the data by 4. As for the mean or median, this method doesn't give a very accurate answer if the data are skewed or contain outliers, but it's a useful rule for well-behaved data.

For example, this rule gives a rough estimate of (42.0–7.8)/4 = 8.55 for the standard deviation of the body fat percentages. The true sample standard deviation is 9.144.

Exercise 3.4:

Estimate the mean and the standard deviation of the per capita incomes listed in Figure 2.6.

Re-scaling

Useful rules exist for computing numerical summaries after re-scaling a set of data. Unlike the approximations described above, these rules give exact results, and may be described as follows.

a. If a constant is added to every measurement in a sample, the same constant is added to the mean, median, and percentiles, but the standard deviation, inter-quartile range, and range are all unchanged.
b. If every measurement in a sample is multiplied by a constant, the same constant multiplies all numerical summaries, including the mean, median, percentiles, midspread, and standard deviation.

These rules may be incorporated into a general formula as follows. If the sample is denoted as $\boldsymbol{y} = (y_1, y_2, \ldots.., y_n)$, then

$$\text{Mean}\,(a\times\boldsymbol{y} + \text{b}) = a\times \text{Mean}\,(\boldsymbol{y}) + \text{b}$$
$$\text{Median}\,(a\times\boldsymbol{y} + \text{b}) = a\times \text{Median}\,(\boldsymbol{y}) + \text{b}$$
$$\text{Percentile}\,(a\times\boldsymbol{y} + \text{b}) = a\times \text{Percentile}\,(\boldsymbol{y}) + \text{b}$$
$$\text{St.Deviation}\,(a\times\boldsymbol{y} + \text{b}) = a\times \text{St.Deviation}\,(\boldsymbol{y})$$
$$\text{Midspread}\,(a\times\boldsymbol{y} + \text{b}) = a\times \text{Midspread}\,(\boldsymbol{y})$$

Rounding

Very often, data are given to more digits of accuracy than you need. For example, the data in Figure 2.6 have five digits, but three digits, or at most four, are sufficient for most purposes. The stem-and-leaf plot of these data given in Figure 2.7 *rounds* these data to three digits.

Consider rounding these data to four digits. The first measurement is 13,679 (for Alabama), which is rounded to 13,680. Note that if the last digit is greater than 5, the second last digit is rounded to the one above. But if the last digit is less than 5, the second last digit is unchanged by the rounding. So the second measurement, 21,173 (for Alaska) gets rounded to 21,170.

When the last digit is 5, a special rule is needed. In this case the best thing to do is to round the second last digit to the nearest *even* digit. This means that the per capita income for Indiana, 16,005, gets rounded to 16,000. Similarly the figure for Oregon (15,785) gets rounded to 15,780, and that for Wyoming (14,135) becomes 14,140.

***Exercise 3.5**:*

Round to integers the pre-race beta-endorphin concentrations of the runners listed in Homework Problem 8 of Chap 2.

Summary

Typical values are used to summarise a set of data.

If just one number is allowed, the mean or the median may be used as a typical value for a set of numerical data, and the proportion of data in a specified (usually the most frequently occurring) category is used to summarise a set of categorical data.

Percentiles are used to define numerical values at specified locations in the distribution of a set of continuously varying data. For example, the median is the 50th percentile, the lower quartile is the 25th percentile, and the upper quartile is the 75th percentile.

The mean or median is a measure of centre. If a measure of spread is required as well, the inter-quartile range and the standard deviation can be used to describe the variability in a set of numerical data.

Both the mean and the standard deviation are quite sensitive to outliers in the data. Outliers do not affect the median and the inter-quartile range very much. For this reason, if outliers are present, the median and inter-quartile range should be used in preference to the mean and the standard deviation.

Methods for getting quick approximations and for rounding data are also given in this chapter.

Homework problems

1. Take two samples each of size 18 from the first 183 records of data listed in Appendix B, the first comprising every tenth person starting with ID 5, the second comprising every tenth person starting with ID 10. For each sample, compute the following summary measures.

   a. the proportion whose preferred drink is an alcoholic drink;

   b. the mean and the standard deviation of the heights;

   c. the mean and the standard deviation of the numbers of hours/week spent watching TV;

   d. the median and the inter-quartile range of the numbers of hours/week spent watching TV.

   Which of the two measures in (c) and (d) more effectively summarises the TV viewing of the sampled students? Explain.

2. For the sulphur dioxide concentrations in Homework Problem 1 of Chapter 2, compute

   a. the median and inter-quartile range, using the stem-and-leaf plot;

   b. the mean and the standard deviation, using your calculator.

3. A *five-number summary* of a sample of numerical data comprises the minimum, lower quartile, median, upper quartile, and maximum of the sample. For the three samples given in Homework Problem 4 of Chapter 2, make a table containing the five-number summaries.

4. Taking the mean of the logarithms of a set of numbers, and then applying the reverse of the logarithm transformation to this mean, gives the *geometric mean* of a set of numbers. For the three sets of data given in Homework Problem 8 of Chapter 2, compute their geometric means and compare them with the medians of the 3 samples.

5. For the psychologists listed in Homework Problem 10 of Chapter 2, find those corresponding to (a) the 20th percentile, and (b) the 95th percentile of the number of citations.

6. Construct a frequency distribution of the lengths of words in the first paragraph of this chapter (beginning with *In the preceding chapter* and ending with *into categories*).

   a. Using the frequency distribution, determine the median, quartiles, range, and midspread of this sample.
   b. Compute the mean and the standard deviation of the sample.
   c. Which measures of location and spread do you think are the most appropriate for summarising these data? Explain.
   d. Can you think of a reason why a researcher would be interested in analysing the distribution of word lengths? Comment.

7. Consider the following set of digits:

   1 10 3 6 1 8 2 4 4 0 10 9 6 6 3 7 5 5 9.

   a. Without doing any sums, determine:
      i. is the mean closer to 1, 5 or 10? Why?
      ii. is the standard deviation closer to 3 or 6? Why?
   b. What happens to the mean and the standard deviation when
      i. 5 is added to each value?
      ii. each value is multiplied by 3?
      iii. each value has its sign reversed?
   c. Find the precise mean, median, range, and standard deviation of the given data.

8. Consider the data for the students listed in Appendix B. Referring to the responses to the question about workload for the first 51 students listed, construct a table giving the sample size, mean and standard deviation of the students from each of the six Australian states and territories represented.

9. From the 51 Australian-born students in Problem 8, compute the 10th and 90th percentiles of the distances travelled to the University.

10. Make a table summarising the drink preferences of the first 186 students listed in Appendix B, by first determining the five most popular drink preferences among these students, and then giving the percentages of students preferring these drinks.

THE NORMAL DISTRIBUTION

CHAPTER 4

4 The Normal Distribution

In this chapter we consider the shapes of population distributions of continuous data. There is a particular distribution that has a central role in Statistics. This distribution is known as the *normal* distribution.

We begin with the idea of smoothing a histogram, in an attempt to uncover the shape of the underlying population distribution. To see where the normal distribution fits in, it is useful to look at other shapes of distributions of commonly occurring data. These include various kinds of symmetric distributions as well as skewed distributions.

The normal distribution may be used as a model for calculating the probability of getting an unusual observation, and the method for doing this is described in this chapter. It is based on the use of the *standardised* normal distribution. In the final section of this chapter a method for assessing the normality assumption, called the *normal scores plot,* is described.

Smoothing histograms

Consider a population of measurements of continuously varying data. Some measurements have distributions that are quite symmetric, whereas others are more or less skewed.

158	170	172	174	175	177	178	180	180	181	184	187
160	170	172	174	175	178	178	180	180	181	185	187
160	170	172	174	175	178	179	180	180	182	185	188
161	170	172	174	175	178	179	180	180	182	185	188
162	170	173	174	175	178	179	180	180	182	185	188
165	170	173	175	175	178	179	180	180	182	185	189
165	170	173	175	176	178	179	180	180	182	185	190
165	170	173	175	176	178	179	180	180	182	185	190
165	170	173	175	176	178	179	180	180	182	185	190
166	170	173	175	176	178	180	180	180	183	185	194
167	170	173	175	176	178	180	180	180	183	185	196
168	171	173	175	176	178	180	180	180	183	185	
170	171	174	175	177	178	180	180	180	183	186	
170	171	174	175	177	178	180	180	180	183	187	
170	171	174	175	177	178	180	180	180	183	187	

Figure 4.1: Heights of 176 Australian-resident male students

Figure 4.1 lists, in order, the heights in centimetres of 176 men who attended a Statistics lecture at Macquarie University on a fine May day.

Figure 4.2 gives a histogram of these data, using data cells (bins) of width 4 cm to group the data, with a smooth symmetric curve superimposed on it. This underlying curve might represent the shape of the distribution of heights in the target population.

Do you believe that the underlying curve could represent the distribution of heights in the target population? Before you answer this question, look at Figure 4.3. This is a histogram of precisely the same data.

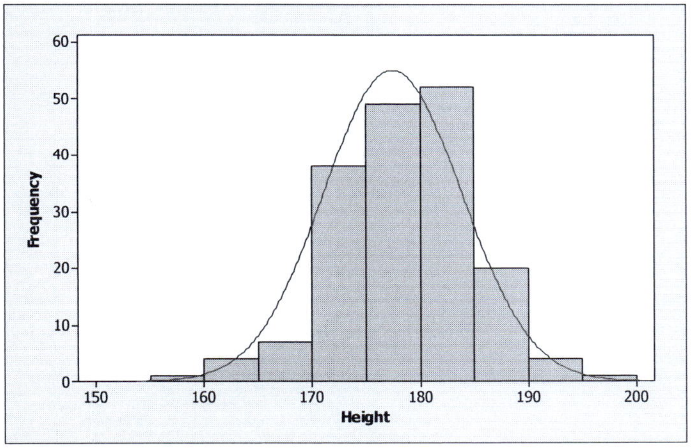

Figure 4.2: Histogram of heights of 176 male students

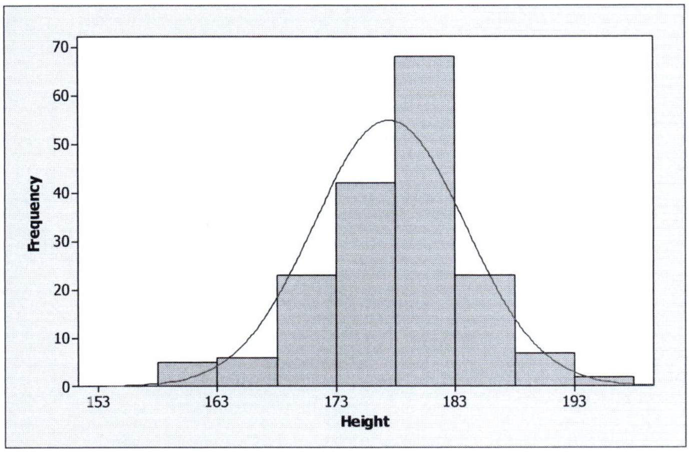

Figure 4.3: Alternative histogram of heights of male students

The only difference in the graphs is that in Figure 4.2 the histogram bins are 155-<160, 160-<165, 165-<170, …, 195-<200, whereas in Figure 4.3 the histogram bins are 158-<163, 163-<168, …, 193-<198. Note that the bin with the largest count in the first histogram (180-185) contains 52 data values, while the bin with the largest count in the second histogram (178-183) has 68 data values.

The message is that the shape of a histogram is affected quite a lot by the choice of bins for grouping the data. Given this fact, it is quite possible that the distribution of heights in the target population could be described by the smooth symmetric curve?

Consider another example. Figure 4.4 gives a histogram for the distribution of the responses of all 426 students attending the Statistics lecture to a question about the workload of the unit. The responses were coded on a scale from 1 (light) to 10 (heavy). These observations were obtained by getting each student to mark a cross on a linear scale divided into ten equal segments, so that the data are integers from 1 to 10.

In this case the underlying population distribution is clearly skewed to the left because it has a long tail to the left. However, the superimposed curve may be too simple in failing to accommodate a peak in the middle of the distribution.

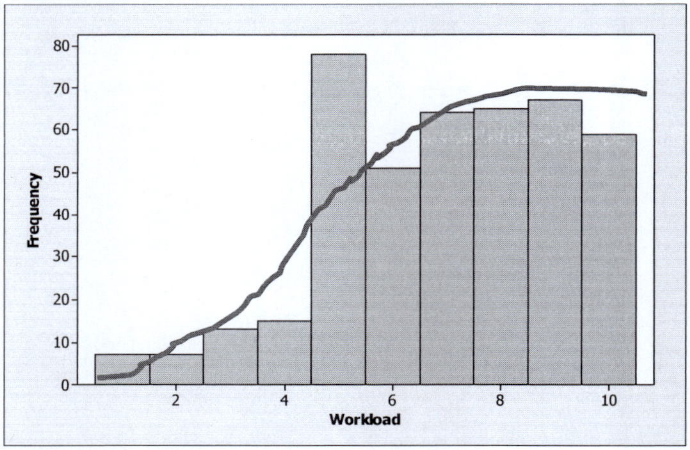

Figure 4.4: Histogram of responses of 426 students

Figure 4.5 shows a histogram of some data that are skewed in the other direction. These are the distances in kilometres that 421 students said they travelled to get to the university, ranging from 0 (for students living on campus) to 140 kilometres. The underlying smooth curve could represent the distribution of travel distances in a larger population such as the population of all students enrolled at the university.

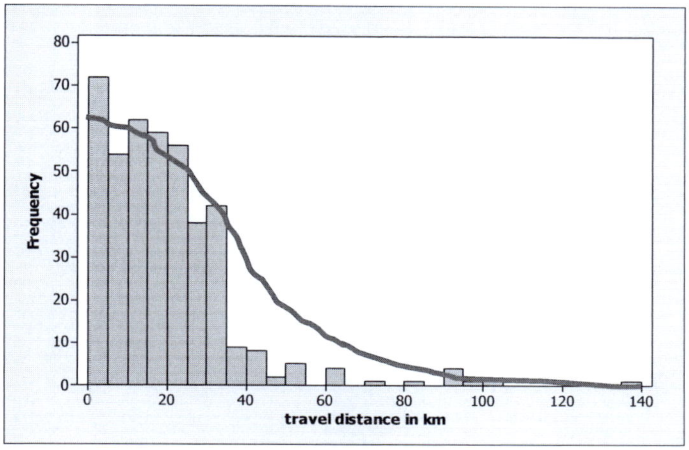

Figure 4.5: Histogram of distances travelled to University for 424 students

A distribution may be also uniform, or rectangular in shape. Figure 4.6 shows the distribution of the month of birth of the students. You might assume that the likelihood of being born on each day of the year is the same. However, the underlying population distribution is not quite uniform because different months have different numbers of days, and the curve connecting the points superimposed on the histogram in Figure 4.6 takes this monthly variation into account.

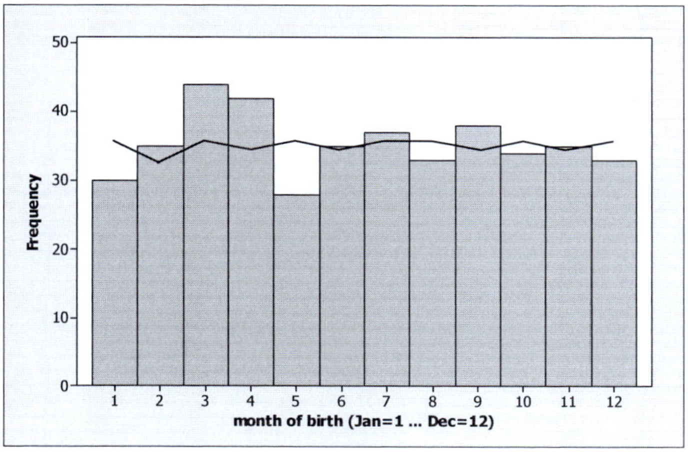

Figure 4.6: Histogram of month of birth of 424 students

As you can see, the curve doesn't fit the histogram very well. In fact, as demographers know, it is not realistic to assume that the birth rate is the same for every day of the year.

Shapes of population distributions

It is useful to think of population distributions as smooth curves, or models that serve as approximations to the truth. For example, even though the curve connecting the rectangles shown in Figure 4.6 may be a better approximation to the truth, you may prefer an even simpler model in which the probability of being born in each month is the same. (This would correspond to an exactly uniform population distribution.)

A uniform (or rectangular) distribution is a special kind of *symmetric* distribution. Figure 4.7 shows four different symmetric distributions, including the rectangular distribution in the lower left. These distributions differ in the extent to which the data are concentrated in the centre.

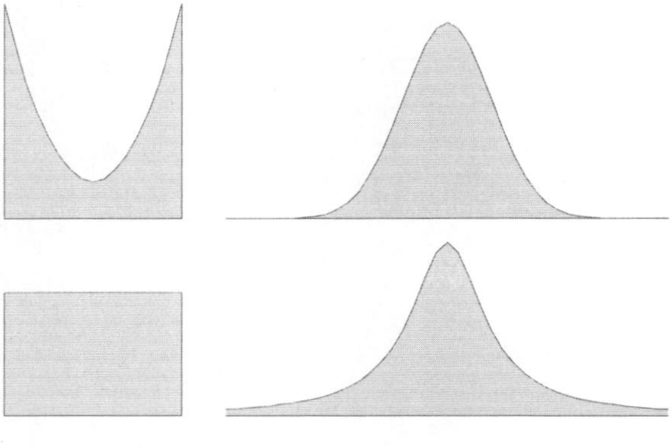

Figure 4.7: Shapes of some symmetric population distributions

The U-shaped curve in the top left of Figure 4.6 could be used to represent symmetrically distributed data which tend to be concentrated at one extreme or the other, such as the number of hours of sunshine in a day. If most days are either almost completely sunny or almost completely cloudy, this model might be a reasonable representation.

The two distributions on the right of Figure 4.7 represent data that tend to be concentrated in the middle of the range. These distributions differ from those on the left of Figure 4.7 in the sense that they have *tails* which extend indefinitely in both directions. The lower curve is more peaked, and longer tailed, than the upper curve.

The curve in the top right of Figure 4.7 has a special role in Statistics, and is known as the *normal* distribution. As you will see in the next chapter, it arises as the limiting distribution of measurements obtained by adding up or averaging an increasing number of components. Since data often arise in practice through some kind of accumulation, it is not unreasonable to use the normal distribution as a model for naturally occurring populations. For example, examination scores are obtained by adding up scores from different sections and thus tend to be normally distributed.

The normal distribution might also be regarded as an *ideal* shape for a distribution, in the sense that, among symmetric distributions with unlimited range, it has a perfect shape: not too long-tailed, and not too peaked in the middle.

This statement is very subjective. But don't forget that Statistics is an art as well as a science.

Figure 4.8 shows two population distributions with shorter and longer tails, respectively, than the normal distribution, with the normal distribution superimposed on each curve.

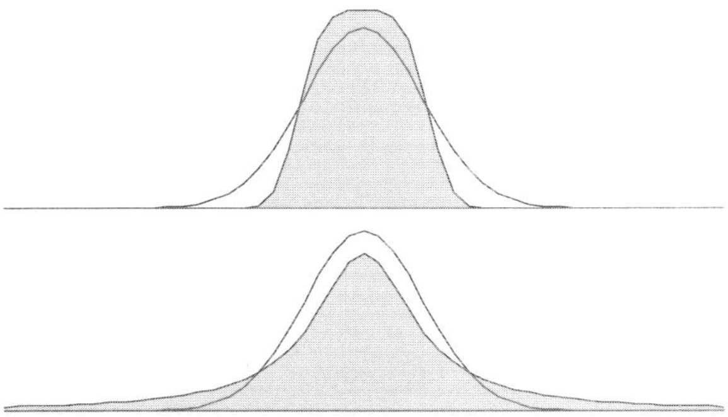

Figure 4.8: Distributions shorter tailed (upper) and longer tailed (lower) compared to the normal distribution

Properties of the normal distribution

A normal distribution is described by two parameters, μ (the centre of the distribution) and σ (a measure of spread). The distribution is symmetric, so μ is the point where the maximum value occurs, and this also corresponds to the population mean. As shown in Figure 4.9, the interval of length 2σ centered at μ contains approximately 68% of the distribution, and the interval of length 4σ centered at μ contains approximately 95% of the distribution (95.4%, to be more exact).

It may be shown that the spread parameter σ arising in a normal distribution is the standard deviation of the distribution. In relation to the curve shown in Figure 4.9, σ may be interpreted at the distance from the centre to a point corresponding to the steepest point on the curve. Thus if you imagine that the normal curve is a cross section of a mountain you are climbing (starting on the left side), the gradient gradually increases until you reach the point $\mu-\sigma$, and then decreases until you reach the top.

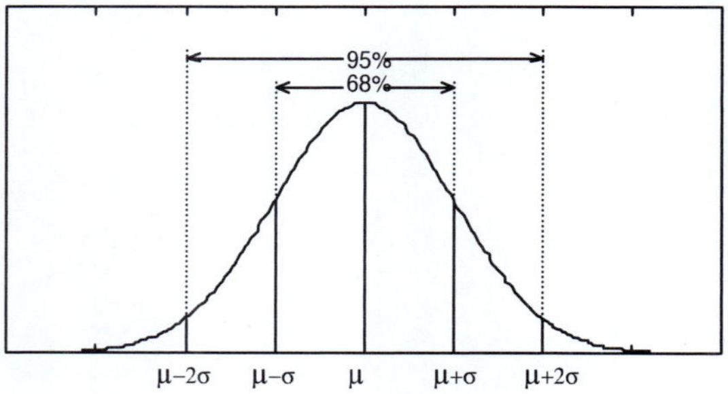

Figure 4.9: Normal distribution with parameters μ and σ

A normal distribution is *standardised* if the central parameter μ is 0 and the spread parameter σ is 1.

If you are prepared to assume that your data comprise a sample from a normal distribution with known parameters μ and σ, you can calculate various probabilities of interest based on this assumption. Suppose, for example, you are a male with a height of 185 cm, and you would like to know the proportion of taller men in your population. This proportion may be regarded as the probability that a male, sampled at random from this population, has a height exceeding 185 cm.

There are two ways of tackling this problem. One way is to take a sample of men's heights from your target population, simply compute the proportion of members in this sample exceeding 185 cm, and use this proportion as an estimate of the probability of interest. For the data listed in Figure 4.1, there are 14 values out of a total sample size of 176 (8%) exceeding 185, so the estimate 0.08 is obtained for the probability.

***Exercise 4.1**:*

Respondents tend to give round numbers (170, 171, 172, 173, etc.) for their height. (This is called 'data heaping'.) So it is likely that only about half of the men in the sample who gave their height as 185 cm were really taller than 185 cm. Using this information, revise the estimate of the proportion exceeding 185 cm.

The other approach involves assuming a known distribution for heights in the population, and then making a calculation using the properties of this distribution. While this approach is less direct, it may be preferable to the direct approach in that it gives an answer that depends only on the assumed population, and not on the particular sample of data. If the population has a normal distribution, only two parameters are needed: the population mean, μ, and the population standard deviation, σ.

The second method is thus less empirical than the first, since it is based on a theoretical model (that is, an assumed normal distribution) rather than on the data in a sample. However if there are good reasons to believe in a theoretical model, it is preferable to make use of such information rather than relying solely on just a sample of data.

The method of computing probabilities based on the normal distribution is described in the next section. Note that the method is valid if, and only if, the population distribution is normal, so it is important to check this assumption. Obviously, it would not be valid for a uniform distribution.

Computing probabilities using the normal curve

As mentioned above, a normal distribution is standardised if the central parameter μ is 0 and the spread parameter σ is 1. Data values in a population having a standardised normal distribution are usually denoted by the symbol z, and are called z-scores. Table A1 in Appendix A gives *tail areas* under this normal curve, a tail area being defined as the area under the curve beyond a specified z score. Using this table you will find, for example, that the area under the curve beyond the point $z = 1$ is

0.1587. Similarly, the tail area corresponding to $z = 0.5$ is 0.3085, while that for $z = 2$ is 0.0228.

Since the normal distribution is symmetric, tail areas need to be tabulated only for non-negative values of z. Using Table A1 you can compute the probability that a sampled measurement from any normal population is greater than some specified value. To do this, two steps are involved:

Step 1: Convert the specified value to a z score.

Step 2: Find the tail area based on the standardised normal distribution.

The first step involves subtracting the population mean from the specified value and dividing this residual by the population standard deviation. So if y is the specified value, z is given by the equation

$$z = \frac{y - \mu}{\sigma} \tag{4.1}$$

Note that z is the number of standard deviations that y is above μ.

The second step is straightforward if z is nonnegative. You just find the entry in the table corresponding to the calculated value of z. Note that the table assumes that z is specified to two decimal places, so you will need to round the z-score to this degree of accuracy.

If z is negative, you need to first compute the tail area corresponding to $|z|$, and then subtract the result from 1.

As an illustration, suppose you are given that IQ scores in a population are normally distributed with mean 100 and standard deviation 15. What proportion of scores is greater than 125?

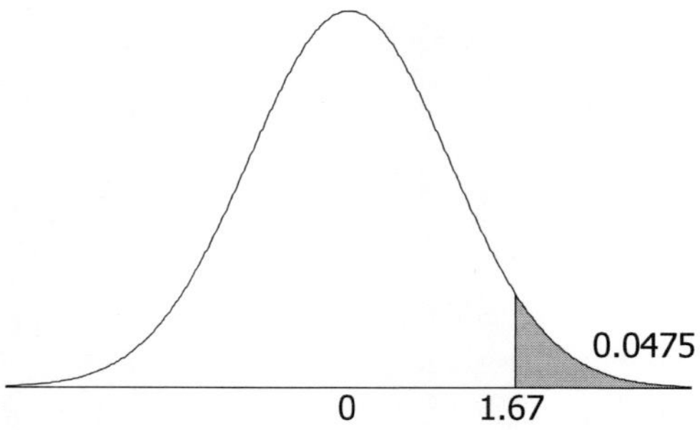

Figure 4.10: Proportion of IQ scores exceeding 125, computed from the standardised normal distribution

To work this out, first compute the z score, that is, the specified value (125) needs to be converted to a z-score using Equation 4.1. Substituting $\mu = 100$ and $\sigma = 15$, this gives

$$z = \frac{y-\mu}{\sigma} = \frac{125-100}{15} = 1.667$$

Using Table A1, the tail area corresponding to $z = 1.67$ is 0.0475, as illustrated in Figure 4.10. So 4.75% of IQ scores exceed 125.

As a further example, suppose that you are given that the heights of male students in the target population of interest are normally distributed with mean $\mu = 177$ cm and standard deviation $\sigma = 6$ cm. To calculate the proportion in the population exceeding 185 cm, proceed as follows:

As before, the specified value, 185, needs to be converted to a z score. Substituting $\mu = 177$ and $\sigma = 6$ into Equation 4.1, this gives

$$z = \frac{y-\mu}{\sigma} = \frac{185-177}{6} = 1.33.$$

The second step involves finding the tail area corresponding to this z score. From Table A1, when z is 1.33 the tail area is 0.092.

So the conclusion is that, based on the normal distribution, 9.2% of the population exceed 185 cm. This result may be compared with the proportion obtained earlier using the direct method, namely, 8% (or less, allowing for the data heaping).

The next example involves a negative z score. Suppose that trees in a *pinus radiata* plantation have a mean height of 16 metres and a standard deviation of 1.25 metres. Assuming they are normally distributed, find the proportion having heights greater than 14 metres.

Substituting $\mu = 16$ and $\sigma = 1.25$ into Equation 4.1, when y is 14, we find that z is $(14-16)/1.25 = -1.6$.

Since z is negative, determine the tail area corresponding to $|z| = 1.6$, namely, 0.055, and subtract this from 1 to obtain the result 0.945. Conclude that 94.5% of trees exceed 14 metres in height, as shown in Figure 4.11.

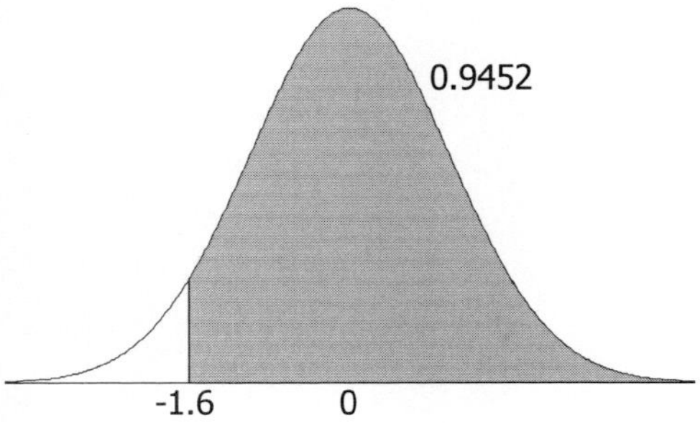

Figure 4.11: Proportion of trees higher than 14 metres, computed from the standardised normal distribution

Exercise 4.2:

A student is informed that that her IQ is 127. What proportion of students in the population has higher scores? (Assume that the population has mean 100 and standard deviation 15.)

Exercise 4.3:

The maximum range (R+S) of an episode of electrocardiograph pulses for healthy males aged 20-40 years is believed to follow a normal distribution with mean 25 mm and standard deviation 9 mm. A measurement of 45 mm is obtained for a 30 year-old schoolteacher. How unusual is this?

Computing data values from tail probabilities

The method outlined above shows how to compute tail probabilities based on data values. It is useful to be able to start from the other end and calculate data values corresponding to specified tail probabilities.

For example, in the case of the IQ scores, you may wish to know what is the IQ score that separates the top 10% of students from the rest.

To solve this kind of inverse problem, proceed as follows:

Step 1: Use the standardised normal distribution to determine the z-score corresponding to the specified tail area. Note that z is positive if the probability corresponds to the upper tail, and z is negative if this probability refers to the area in the lower tail (or if the probability of the value being exceeded is greater than 0.5).

Step 2: Compute the value corresponding to the z score, using the formula

$$y = \mu + \sigma z \tag{4.2}$$

Note that Equation 4.2 is just Equation 4.1, inverted to give y in terms of z rather than z in terms of y.

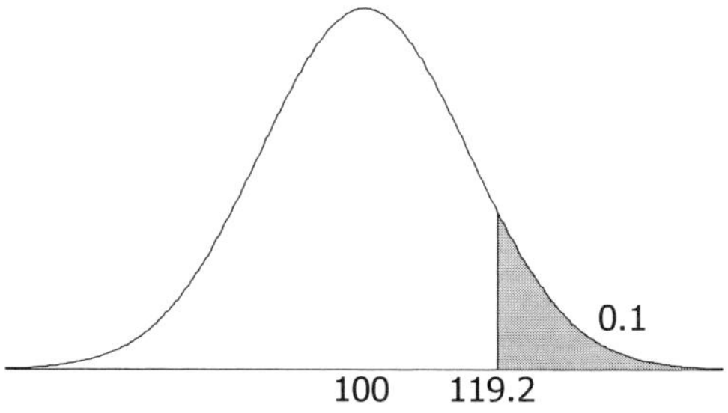

Figure 4.12: The IQ separating the top 10% when μ is 100 and σ is 15

Applying this method to determine the IQ separating the top 10%, Step 1 gives the z score 1.28, and, substituting $\mu = 100$ and $\sigma = 15$, Step 2 then gives $y = 100 + 15 \times 1.28 = 119.2$. Conclude that an IQ of 119 separates the top 10% of students (as shown in Figure 4.12).

Exercise 4.4:

Suppose that the distribution of Graduate Record Examination scores for admission to US graduate schools in Mathematics is normally distributed with a mean of 560 and a standard deviation of 120. A Mathematics professor considers that only 8% of candidates who take this test are capable of undertaking graduate study in Mathematics. What score separates this group?

Exercise 4.5:

A contractor supplies building blocks with an average weight of 92 kg and a standard deviation of 16 kg. A customer agrees to buy one third of these blocks, but wants the largest stones available. What is the minimum weight of these larger stones, assuming that the weights are normally distributed?

The probability of an unusual event

Statistics involves the assessment of probabilities of remarkable or unusual events. 'Unusual' means extreme, that is, an observed measurement is unusual if it is extremely high or extremely low. For example, a maximum temperature of 40 degrees Celsius in Sydney in January is unusual because it is more than 10 degrees above the average. A maximum of 12 degrees is also unusual because it is more than 10 degrees below the average.

If the maximum temperature in Sydney on days in January is assumed to have a normal distribution with known mean and standard deviation, you can make a precise statement about how unusual an observed value is. If the value was higher than average, you could use the upper tail area of the normal distribution as a precise measure of the probability (or *improbability*) associated with the event, and if the value were below average you could use the lower tail area. But you could also use the combined probability in both tails, on the grounds that the event would be considered just as remarkable if it had been as extreme in the other direction.

Since the normal distribution is symmetric, the probability in the two tails is obtained simply by doubling the probability in the upper tail. Consequently the total probability associated with an unusual event is obtained by calculating the probability of getting an observation larger in absolute value, and doubling this probability.

For example, suppose that the maximum temperature on January days in Sydney is normally distributed with a mean of 27 degrees Celsius and a standard deviation of 5. On a particular day the temperature is 39. How unusual is this?

Following the method described in the preceding section, the z score is determined as $(39–27)/5 = 2.40$. From Table A1, the upper tail area is 0.0082, and the corresponding area in the two tails is 0.0164. You may conclude that the probability associated with this extreme temperature is approximately 0.016. This is the probability of getting a temperature of 39 or more, or of 15 or less. The justification is that if you had observed a maximum temperature of only 15 degrees, you would have considered it just as remarkable, because a temperature of 15 is the same amount below the average as 39 is above it.

Assessing the normality assumption

The probability calculations described in the preceding section assume that the population is normal, and are thus not valid otherwise.

You can assess the normality assumption to some extent by looking at a histogram of the data. For example, looking at the histograms in Figures 4.2 and 4.3, you might conclude that the population of men's heights could be normally distributed, but you would conclude from Figures 4.4, 4.5 and 4.6 that none of these population distributions is normal.

In many cases it is more difficult to assess normality from a histogram. Skewness is fairly easy to see in a histogram, but if the histogram is approximately symmetric it is often very difficult to assess normality.

One of the best ways to tell if data could have arisen as a sample from a normal population is to look at a *normal scores plot* of the data. This is a scatter plot obtained by plotting the observations sorted in ascending order on the horizontal axis and against these plotting the corresponding expected values based on a standardised normal distribution. In other words, a normal scores plot is a plot of the points (x_i, y_i), $(i = 1, 2, \ldots, n)$, where x_i is the ith smallest observation and y_i is the corresponding expected value from a standardised normal distribution. These expected values are called *normal scores*, and are based on probabilities associated with the normal distribution. For example, if the sample size n is 10, the normal scores are

−1.335 −0.908 −0.605 −0.349 −0.114 0.114 0.349 0.605 0.908 1.335.

The normal scores may be calculated from a table of standardised normal curve areas by finding the z-scores corresponding to the tail areas $1/(n+1)$, $2/(n+1)$, $3/(n+1)$, etc., where n is the sample size. So for $n=10$, $1/(n+1)$ is $1/11 = 0.0909$, and the z-score corresponding to this tail area is 1.335. The next largest normal score is that corresponding to the tail area $2/(n+1) = 0.1818$, that is, 0.908.

Exercise 4.6:

Compute the normal scores for a sample of size 4.

If the data comprise a sample from a normal distribution, the normal scores plot should show a linear trend. Figure 4.13 shows a normal scores plot for the sample of male student heights listed in Figure 4.1.

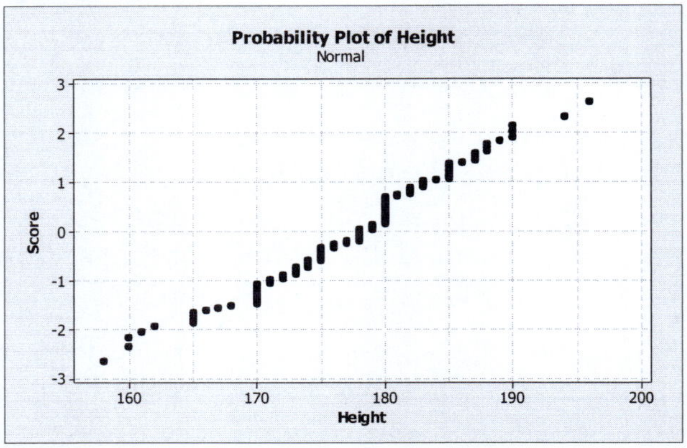

Figure 4.13: Normal scores plot of sample of 176 male student heights

As a further example, Figure 4.14 shows the normal scores plot for the sample of 421 distances travelled, for which the histogram is shown in Figure 4.5. In this case there is a pronounced curvature in the points plotted indicating that the normality assumption is clearly untenable.

It is useful to be able to assess how much non-linearity can be tolerated in a normal scores plot before abandoning the normality assumption. Figure 4.15 shows a normal scores plot for a sample of 50 from a normal distribution with mean 177 and standard deviation 6.5 (similar to the male heights) generated using a computer random number generator. This indicates a typical normal scores plot when the data really come from a normally distributed population.

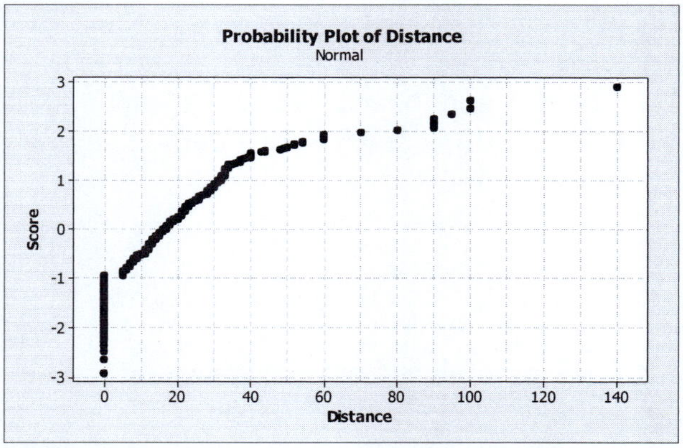

Figure 4.14: Normal scores plot of sample of 421 distances travelled

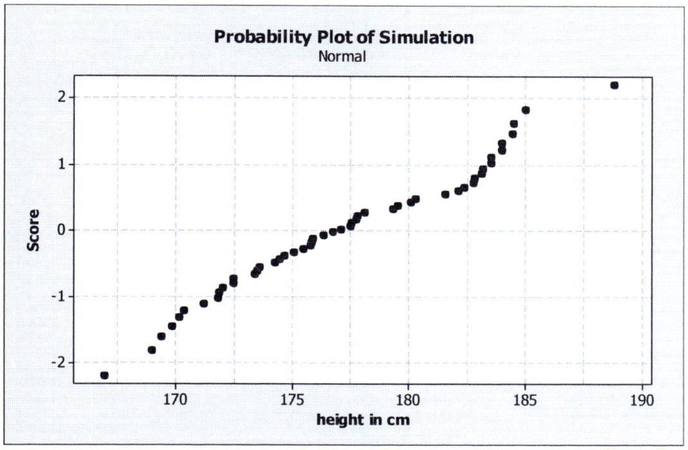

Figure 4.15: Normal scores plot of a simulated sample of size 50 from a normal distribution with mean 177 and standard deviation 6.5

Transforming data to achieve normality

If a histogram is skewed there is evidence that the data in the population are not normally distributed. However, if data are positively skewed it is possible that the normality assumption is reasonable after the data have been transformed in some way, such as by taking logarithms. In fact it is often preferable to transform the data to make their distribution more symmetric before undertaking statistical analysis.

Consider the distances travelled by the students. If the distances are transformed by adding 1 and then taking logarithms to base 10, the normal scores plot is shown in Figure 4.16. (The reason for adding 1 before taking logarithms is because some of the distances are 0, and you can't take the logarithm of 0. But if 1 is added, a distance of 0 remains 0 after the transformation).

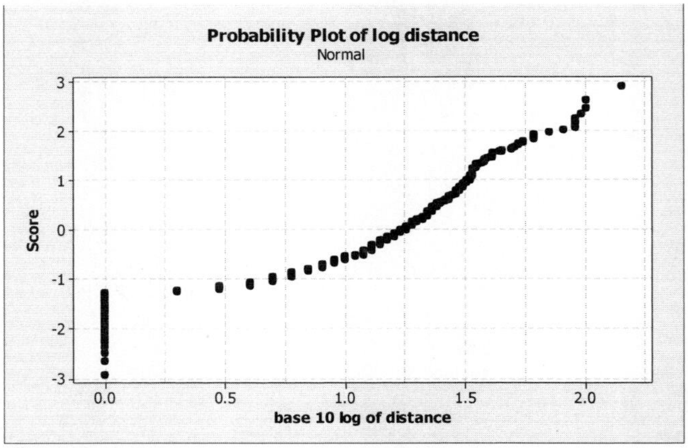

Figure 4.16: Normal scores plot of distances travelled after applying the data transformation $y = log_{10}(x+1)$

Comparing the normal scores plots of the raw data and the log-transformed data (Figure 4.14 and 4.16), the zeroes stand out, indicating the data are not purely continuous. If you wish to focus on the continuous part of the distribution, you could exclude the 44 students who travelled zero distance, and examine the travel distances for the remaining 377.

Figure 4.17 shows the resulting normal scores plot, which does not show any evidence of non-normality. It would thus be reasonable to use the normal distribution as a model for the *logarithms* of the distances travelled by the students who travelled 1 km or more.

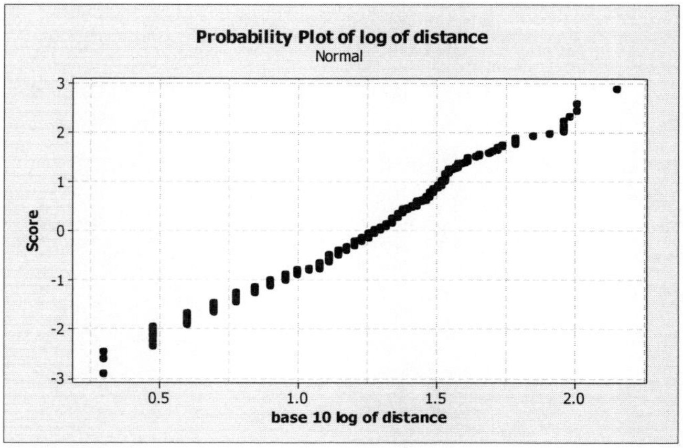

Figure 4.17: Normal scores plot of 377 distances travelled after applying the data transformation $y = log_{10}(x+1)$ and excluding distances of 0

Summary

In this chapter we began with the idea of smoothing a histogram, aiming to uncover the shape of the underlying population distribution. You saw how histograms of the same data differ, depending on the choice of bins for grouping the data to graph the histogram. And you saw how smoothed histograms have different shapes: symmetric or skewed, short-tailed or long-tailed, or even bimodal.

The normal distribution arises as the *ideal* shape for a distribution: not too stubby, but not too stretched in the tails. The normal distribution has two defining parameters, the mean (μ) and the standard deviation (σ). Approximately 68% of the area under a normal distribution is between the values $\mu - \sigma$ and $\mu + \sigma$, while approximately 95% of the area under this curve is between $\mu - 2\sigma$ and $\mu + 2\sigma$. The *standardised* normal curve has mean $\mu = 0$ and standard deviation $\sigma = 1$.

The normal distribution is used as a model for computing the probabilities of extreme events. For example, if you assume that some data are sampled from a normal population with known mean and standard deviation, you can compute the probability that a sampled observation is greater than (or less than) a given quantity. This method uses a table of the standardised normal distribution to compute *tail probabilities*.

It is also possible to compute the value corresponding to a specified tail probability, simply by reversing the procedure.

The probability of an unusual event may be defined as the sum of two tail probabilities, or twice the probability associated with a single tail.

These methods are valid only if it is reasonable to assume that the data come from a normal population. To assess this assumption, you examine a *normal scores plot*, which should show a linear trend if the population is normal.

If the normality assumption is not valid, it may be possible to obtain normally distributed data by making a *transformation* of the data, such as taking logarithms.

Homework problems

1. A vending machine dispenses soft drink into a cup that holds at most 175 ml. The volume dispensed is normally distributed with mean 150 ml and standard deviation 8 ml.

   a. What is the probability that the cup will overflow?

   b. How likely is it that a customer will receive less than 165 ml?

   c. If customers tend to complain when the dispensed volume falls below 135 ml, how frequently are such complaints likely to occur?

2. Salaries for junior executives at *The Firm* are normally distributed with mean $45,000 and standard deviation $3000.

   a. You have 'arrived' if you are in the top 15%. What salary does this correspond to?
   b. The top 25% get keys to the executive washroom. Werner earns $41,500. Does he have a key?
   c. Due to a recession the bottom 10% is to be let go. What salary determines this chop point?
   d. The top 20% eat lunch at *Madame's*. The bottom 30% brings their own bag lunch. The remaining 50% eat at the cafeteria. What salaries separate these groups?

3. The IQs of students in the graduating classes of high schools in *Dum* ***Dum*** **are normally distributed with a mean of 104 and a standard deviation of 10. The Education Authorities divide these students into** three groups as follows:

   a. those with IQs below 100, who are sent back to school for another year;

   b. those with IQs between 100 and 110, who join the army;

   c. those with IQs above 110, who go to university.

   There are 130 students in a particular year. How many are expected in each group?

4. A confectionery manufacturer has a policy that not more than 1.5% of the output of its *verry ripe* bars is to be rejected due to failure to meet the desired weight specification of 40 grams. It is known that *verry ripes* produced in the factory have a mean weight of 40 grams with a standard deviation of 800 milligrams. What maximum and minimum weights should be specified so as to comply minimally with company specifications? (Assume that *verry ripes* have normally distributed weights.)

5. For a normal distribution, calculate the proportion of the distribution that is more than one inter-quartile range beyond the quartiles.

6. Use your computer to generate a personalised sample of 500 random numbers from a normal distribution with population mean 10 and standard deviation 2. Compute the mean (m) and standard deviation (s) of this sample. Now determine the proportion of sample values between $m-s$ and $m+s$. Repeat this exercise after (a) squaring, and (b) taking square roots of the data in the original sample. Summarise and explain your conclusions.

7. According to the rules of tennis: 'The ball shall be more than 56.7 grams and less than 58.5 grams in weight. The ball shall have a bounce of more than 135 cm and less than 147 cm when dropped 254 cm upon a concrete base.' A tennis ball manufacturer, knowing that balls will gradually lose their pressure, and hence their bounce, designs their ball according to the following specifications:

   weight: mean = 57.4 grams, st.dev. = 0.4 grams;
   bounce: mean = 143 cm, st.dev. = 3 cm.

   The distribution of weight and bounce of tennis balls is normal.

   a. If a club orders 1000 balls from this manufacturer, how many would you expect to be rejected (if tested) on account of weight?
   b. The C grade club members prefer a lighter ball. How many of the 1000 balls do you expect to weigh between 56.8 and 57.2 grams?
   c. The A grade club members prefer the high bouncing balls and select the 200 (20%) balls that bounce the highest regardless of whether or not they satisfy the bounce test. What is the minimum bounce for balls selected by A grade players?

8. 'Red striped tiger worms and their voracious appetite for kitchen scraps have found popularity as an environmentally friendly answer to Sydney's burgeoning waste problem.' The *Sydney Morning Herald* reported this information on 13 October 1993. It was claimed that each worm could chomp through an average of 1 gram of rubbish a day with a standard deviation of 0.25 grams. The worms are sold in plastic cans containing 1000 worms.
   a. What proportion of worms can eat more than 1.2 grams of waste/day?
   b. What percentage of these worms can eat between 0.8 and 1.05 grams of waste per day?
   c. How many worms in a can would you expect to eat less than 0.8 grams per day?
   d. The top 10% of worms, in terms of their eating capabilities, are to be selected for export. What minimum amount of waste can these worms eat per day?

9. Make a normal scores plot of the sample of 18 body fat percentages listed in Figure 2.1. Do you think the corresponding target population could be normally distributed? Explain.

10. As in Homework Problem 6, use your computer to generate a personalised sample of 100 random numbers from a normal distribution with population mean 10 and standard deviation 2. Produce a normal scores plot of your sample. Repeat this exercise after (a) squaring, and (b) taking square roots of the data in the original sample. Summarise and explain your conclusions.

DISTRIBUTIONS OF MEANS AND PROPORTIONS

CHAPTER 5

5 Distributions of Means and Proportions

The preceding chapter focused on the normal distribution as a model for continuous data. This distribution plays a central role in Statistics, largely because it arises as a result of averaging. In the present chapter we show more precisely how the normal distribution arises as the distribution of sample means.

Sample means are used to obtain more accurate measurements of population parameters of interest. The larger the sample, the more closely it tends to reflect the population from which it is sampled, and in particular the more closely a sample mean reflects the mean of the population. A quantity called the *standard error* provides a measure of this discrepancy, and it is important to know its size and determinants. So a further objective of the present chapter is to provide a definition of the standard error for estimating a population mean.

The final section of this chapter is concerned with the distribution of a sample proportion, and its accuracy, again expressed in terms of a standard error, in estimating the corresponding population proportion. As you will see, a sample proportion is simply a special case of a sample mean, so the methods already developed for sample means may be used.

Averages of uniform random numbers

The University of Hell attracts around 5000 applicants every year, but can only accommodate 1000 new students. For a nonrefundable application fee of $1000, a candidate is assigned a number between 1 and 100, determined by spinning a chocolate wheel containing these numbers. An application is successful if the number obtained is in the range from 41 to 60 inclusive. Assuming that each number from 1 to 100 is equally likely, the chance of acceptance for any particular applicant is thus 20 in 100, or 0.2.

By paying an additional $1000, an applicant can get two numbers, determined by two spins of the chocolate wheel, but only the average of these two numbers is used for admission purposes. If the average is at least 41 and no more than 60, the application is successful.

What is the probability of acceptance for an applicant paying $2000?

To answer this question, you need to know the distribution of the average of two random numbers that are uniformly distributed between 1 and 100. A simulation study may be used to induce the answer.

Figure 5.1 shows the result of such a simulation study, with histograms of a sample of 5000 random integers uniformly distributed between 1 and 100 shown at the top, and the averages of 5000 pairs of independent random integers with the same uniform distribution shown below. There were 1025 'successful' applicants (20.5%) based on the uniform distribution, and 1772 (35.4%) based on the distribution of the averages. So it appears that a student can almost double his or her chances of getting into Hell University by paying the additional $1000.

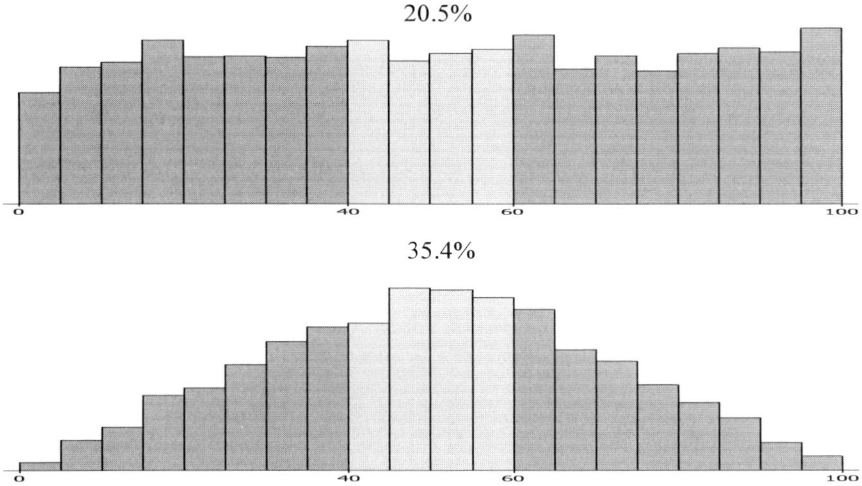

Figure 5.1: Histograms of 5000 uniform random numbers between 1 and 100 (top) and 5000 averages of their pairs (bottom)

It is clear that the averages are more likely to occur in the middle of the range, and the distribution of these averages appears to be approximately triangular in shape.

Going further, consider the distribution of the result of averaging the numbers obtained from more than two spins of the chocolate wheel. Figure 5.2 gives the distributions based on a simulation study of 5000 multiple spins, where the number of components making up the average is 3, 5 and 10, respectively.

You can see that the proportion of results in the range 41-60 increases as the number of terms (n) making up the average increases, from 43.1% when n is 3, 55.0% when n is 5, to 71.3% when n is 10.

Furthermore, the distribution of the averages approximates a normal distribution as n increases.

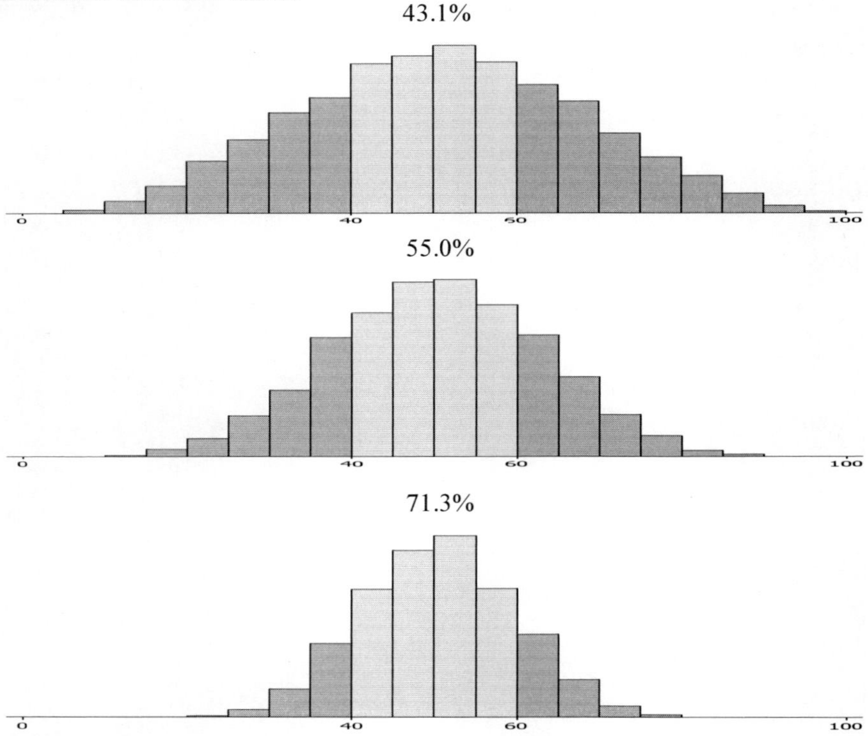

Figure 5.2: Histograms of 5000 averages of sets of n uniform random numbers between 1 &100, for n=3 (top), n=5 (centre), & n=10 (bottom)

Figure 5.3 shows population distributions corresponding to the histograms of the averages shown in Figures 5.1 and 5.2.

This result is a special case of *the central limit theorem.* The central limit theorem states that the distribution of the average of a set of independent random numbers is approximately normally distributed, and that this approximation improves as the number of terms in the average increases.

You can see that the curves on the right hand side of Figure 5.3 are close to the normal curve graphed in Figure 4.9. Even when n is as small as 3, the population distribution (shown in the lower left corner of Figure 5.3) is quite similar to a normal curve.

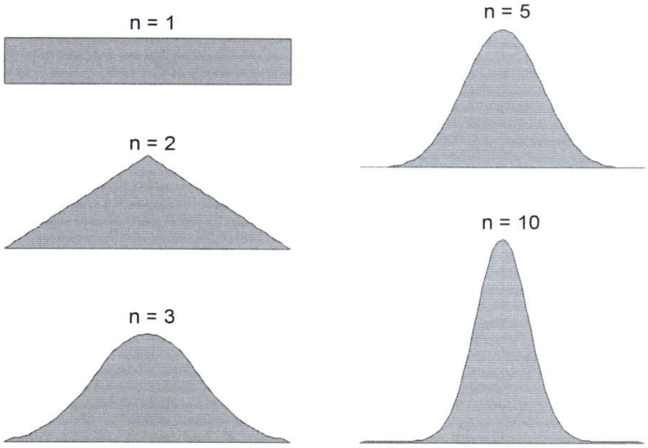

Figure 5.3: Population distributions for averages of sets of n uniform random numbers

Another feature of the distributions shown in Figure 5.3 is their tendency to become more concentrated in the centre as *n* increases. In other words, the more components you average, the more likely it is that you will get a result in the middle of the range. This effect is described in the next section.

The standard deviation of an average

Consider again the simulated samples of 5000 averages. The means and standard deviations of these samples are given in Figure 5.4.

These standard deviations quantify the trend in Figures 5.1 and 5.2, namely, that the spread in the distribution of the averages decreases as the number of components increases.

The standard deviations in Figure 5.4 decrease regularly as *n* increases, but not in direct proportion to *n*. Figure 5.5 shows the standard deviations from Figure 5.4 together with two more columns, containing the result of multiplying the standard deviations by (a) *n* and (b) the square root of *n*.

n	mean	st. deviation
1	50.7	28.6
2	50.1	20.5
3	50.8	16.8
5	50.4	12.8
10	50.4	9.2

Figure 5.4: Means and standard deviations of samples of 5000 averages of n random numbers

n	st.deviation	st. dev.× n	st. dev.×√ n
1	28.6	28.9	28.9
2	20.5	41.0	29.0
3	16.8	50.4	29.1
5	12.8	64.0	28.6
10	9.2	92.0	29.1

Figure 5.5: Model for standard deviation of samples of 5000 averages of n random numbers

You can see from this table that the product of the standard deviation and the square root of n is approximately constant, and the constant is close to 29. Assuming this is correct, you can write down a formula for the standard deviation of a distribution of averages, namely

$$\text{st.dev.} = \frac{const.}{\sqrt{n}} \tag{5.1}$$

If σ denotes the standard deviation of the population of single values, then the constant in Equation 5.1 is σ (since this is the standard deviation when n is 1), so the standard deviation of an average of n components is

$$\text{st.dev.}(\overline{y}) = \frac{\sigma}{\sqrt{n}} \tag{5.2}$$

Equation 5.2 is the formula for the *standard error* of a mean. The term 'standard error' is used for the standard deviation of an estimate of a population parameter based on a sample.

The simulation study demonstrates the validity of Equation 5.2 for uniformly distributed components. However, it turns out that the formula is valid more generally. In fact it applies to means of samples from any distribution having a finite standard deviation. It is possible to prove this result using statistical theory. In the next section we consider a practical application of these ideas.

The distribution of sunshine

The Department of Meteorology records the number of hours of sunshine at various stations, including Observatory Hill, at the southern end of the Sydney Harbour Bridge. The observations for full week (Monday to Sunday) periods in April for 1955, 1956 and 1957 are listed below in Figure 5.6.

Start date	Mon.	Tues.	Wed.	Thurs.	Fri.	Sat.	Sun.
4/4/1955	0.3	1.3	0.0	0.0	0.0	1.3	3.7
11/4/1955	8.9	5.6	7.4	6.0	7.5	1.7	1.2
18/4/1955	5.4	0.5	3.0	8.4	9.1	8.7	8.2
2/4/1956	0.0	1.4	2.3	6.8	9.2	10.0	9.8
9/4/1956	9.2	7.2	9.3	9.4	10.2	2.2	4.9
16/4/1956	3.3	10.5	9.7	7.6	10.3	9.3	8.7
1/4/1957	10.1	7.7	0.0	8.6	9.6	10.0	2.0
8/4/1957	4.1	9.0	8.5	0.1	2.3	2.3	5.3
14/4/1957	1.0	7.0	0.5	0.0	9.2	9.6	10.0

Figure 5.6: Daily hours of sunshine in Sydney in April 1955-57

An observation of 0.0 corresponds to a day when it was cloudy all day. If it is sunny all day the number of hours of sunshine is close to 11, the number of hours from sunrise to sunset in Sydney in late April.

The distribution of the numbers of hours of sunshine, for all complete weeks in April from 1955 to 1997, is shown as a histogram in Figure 5.7. This distribution is based on 903 days (3 weeks in each of 43 years) and is clearly U-shaped, indicating that a day is likely to be almost completely sunny or almost completely cloudy, and less likely to be half-sunny and half-cloudy.

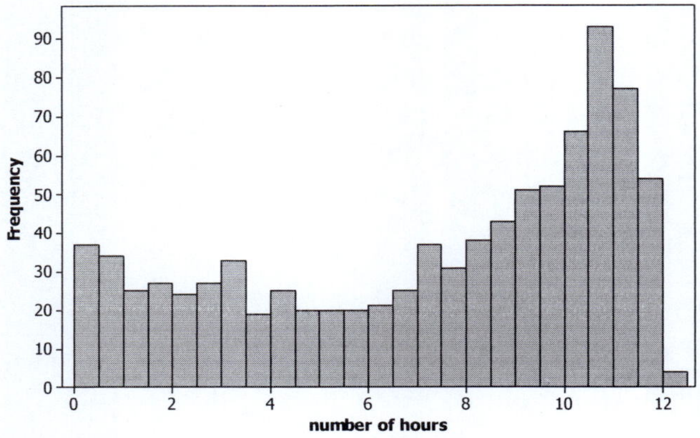

Figure 5.7: Histogram of hours of sunshine in Sydney in April 1955-97

If you must work or attend classes on weekdays and you like to relax during the weekend, you may be more interested in the amount of sunshine on Saturdays and Sundays. You would expect the amount of sunshine on a Saturday or a Sunday to have the same distribution as that on a weekday. But suppose that you are more interested in the *average* number of hours of sunshine per day during the weekend. Figure 5.8 shows histograms of the number of hours of sunshine on weekend days, and the average number of hours of sunshine on weekends. The top histogram is thus based on 258 days (6 weekend days over 43 years), while the bottom is based on 129 weekends (3 weekends over 43 years).

As expected, Figure 5.8 shows that the distribution of the number of hours of sunshine on weekend days is not very different from that for all days, as shown in Figure 5.7. However, the distribution of the averages is more concentrated in the middle. Thus corresponds to the fact that it could be sunny on the Saturday and cloudy on the Sunday, or vice versa, giving a middling amount of sunshine on average.

Consider now a person with an outdoor job, such as a bricklayer or market gardener, who prefers to stay inside during the weekend. This person would be interested in the average number of hours of sunshine during a weekday. Figure 5.9 gives histograms of the number of hours of sunshine on weekdays, and the average number of hours of sunshine on weekdays. The top histogram is thus based on 645 days (15 weekdays over 43 years), while the bottom is based on 129 days (3 weekday weeks over 43 years). Again you can see that the distribution of the averages is more concentrated in the middle of the range.

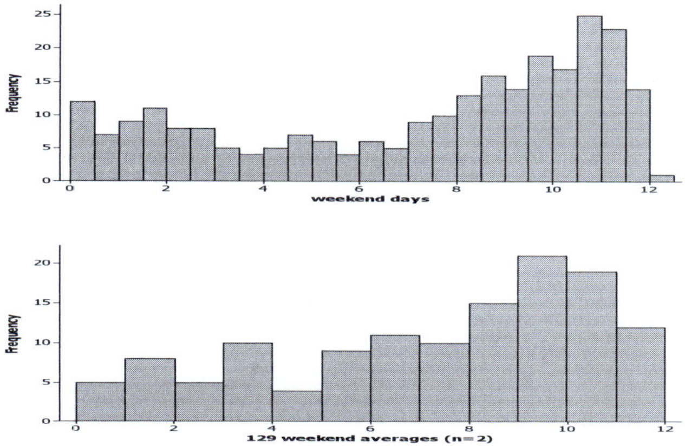

Figure 5.8: Histograms of number of hours of sunshine on weekend days (upper panel) and average number of hours of sunshine (lower panel)

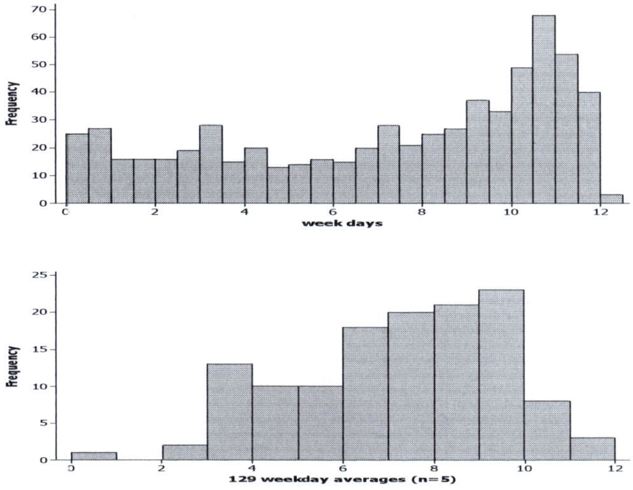

Figure 5.9: Histograms of number of hours of sunshine on week days (upper panel) and average number of hours of sunshine (lower panel)

Finally, Figure 5.10 compares the histograms of the daily totals, the weekend averages, the weekday averages, and the averages for whole weeks. This graph may be compared with Figures 5.1 and 5.2, which give similar results based on simulated data.

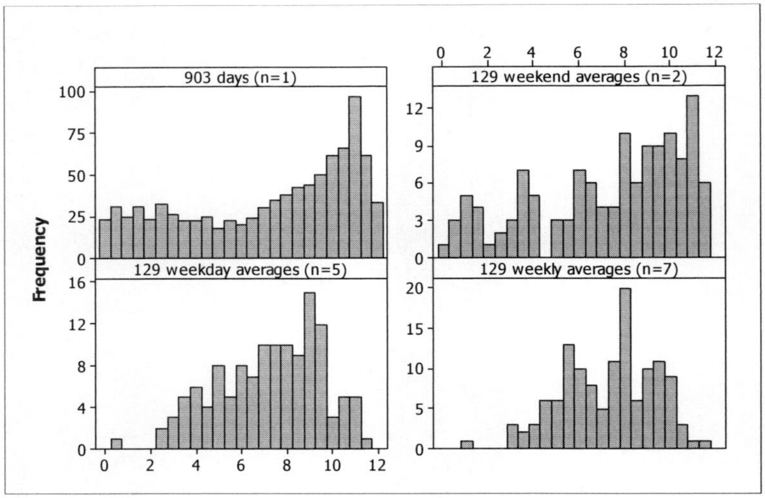

Figure 5.10: Histograms of number of hours of sunshine in April days (upper panel) and averages of 2, 5 and 7 days (lower panels)

Figure 5.11 lists the standard deviations of these samples of averages.

Number of components (n)	Observed st. deviation	Theoretical st.dev ($s/\sqrt{n}$)
1	3.39 (s)	
2	2.08	2.40
5	1.39	1.52
7	1.11	1.28

Figure 5.11: Standard deviations of average number of hours of sunshine for weekends (n=2), workday weeks (n=5), and full weeks (n=7)

According to Equation 5.2, these standard deviations should decrease with the reciprocal square root of n, where n is the number of components being averaged. To check this theory, the values of $s/\sqrt{n}$ (where s = 3.39 is the standard deviation of the sample of individual sunshine measurements) are also given in Figure 5.11. As you can see, the standard deviations of the averages are consistently smaller (by approximately 10%) than the corresponding theoretical values.

The discrepancy between the observed and theoretical values in Figure 5.11 can be explained as follows. The formula giving the standard error of an average (Equation 5.2) is based on the assumption that the components in the average are completely *independent* of each other. However, the numbers of hours of sunshine in successive days are correlated, and thus not independent. If it is sunny on a particular day, it is likely that the next day will also be sunny. Similarly, cloudy days are more likely to be followed by cloudy days than by sunny days. You can see this by looking at the data in Figure 5.6, where there are six cloudy days in a row in the week beginning on the 4th of April 1955, and four sunny days in a row from the 6th of April 1956. With correlated data there is less variation and so the standard deviation is smaller.

Probability calculations involving averages

In the preceding chapter the normal distribution was introduced as a model for an 'ideal' population, and we saw how this model could be used to compute population proportions. For example, if we know that IQ scores are normally distributed with mean 100 and standard deviation 15, we calculate that 4.75% of persons in this population have IQs above 125.

In many situations, you may need to calculate the probability that an *average* of a fixed number of components is greater than some specified value.

Suppose, for example, that you know that the diastolic blood pressure (DBP) for normal adults is normally distributed with mean 90 mm and standard deviation 10 mm. A person has a value of 102 for their DBP, based on the average of three readings from a sphygmomanometer. What is the likelihood of getting a reading as high as this?

To answer the question, it is necessary to know the distribution of an average of three, as shown in the left panel of Figure 5.12.

Since each component has a normal distribution with mean $\mu = 90$ and standard deviation $\sigma = 10$, the average is also normally distributed with mean 90, and standard deviation given by Equation 5.2, that is

$$\text{st.dev.}(\bar{y}) = \frac{\sigma}{\sqrt{n}} = \frac{10}{\sqrt{3}} = 5.77$$

So, using the method outlined in Chapter 4, you need to compute the z score using the formula (Equation 4.1)

$$z = \frac{y - \mu}{\sigma}$$

and then obtain a tail area based on the standardised normal distribution.

Since we are dealing with an average based on n components instead of just a single measurement, the formula in this case is

$$z = \frac{\bar{y} - \mu}{\sigma / \sqrt{n}} \tag{5.3}$$

The z-score for an average diastolic blood pressure of $\bar{y} = 102$, based on a sample of $n = 3$ measurements from a population with mean $\mu = 90$ and standard deviation $\sigma = 10$, is thus

$$z = \frac{102 - 90}{10 / \sqrt{3}} = 2.08$$

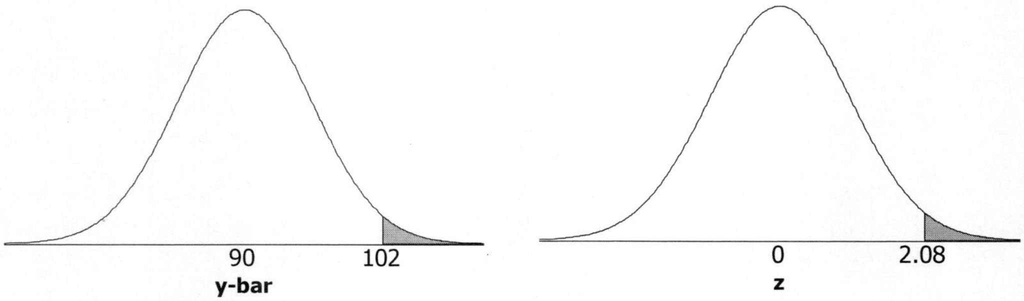

Figure 5.12: Distribution of the average of three diastolic blood pressure readings (left panel) and standardised normal curve (right panel)

Using the standardised normal distribution, the upper tail area, corresponding to the probability of getting a z score greater than this value, is only 0.0188, as shown in the right panel of Figure 5.12. So you may conclude that a diastolic blood pressure of 102, based on three readings, is unusually high.

The following problem provides a further illustration of the method of finding a probability associated with an average. Suppose that a market gardener supplies lettuces to a supermarket with a mean weight of 1 kg and a standard deviation of 100 grams (0.1 kg). The lettuces are packed in boxes of 12. A quality control inspector for the supermarket selects a box at random and finds that it weighs only 11 kg. What is the likelihood of getting such a box weighing 11 kg or less? (Assume the weights of the lettuces are normally distributed.)

First note that the probability that the *total* weight of a box of 12 lettuces is less than 11 kg is the same as the probability that the *average* weight of a lettuce is less than 11/12 kg, or 917 gm. Since the individual lettuce weights are normally distributed with mean $\mu = 1000$ gm and standard deviation $\sigma = 100$ gm, the average is also normally distributed with mean 1000 gm, and standard deviation given by Equation 5.2. The z score is given by Equation 5.3 with $\bar{y} = 917$, $\mu = 1000$, $n = 12$, and $\sigma = 100$, that is

$$z = \frac{\bar{y} - \mu}{\sigma / \sqrt{n}} = \frac{917 - 1000}{100 / \sqrt{12}} = -2.88\,.$$

Based on the standardised normal distribution, the (left) tail area corresponding to this value of z is only 0.0020, as shown in Figure 5.13. You may conclude that a box of lettuces weighing only 11 kg is quite an unlikely occurrence.

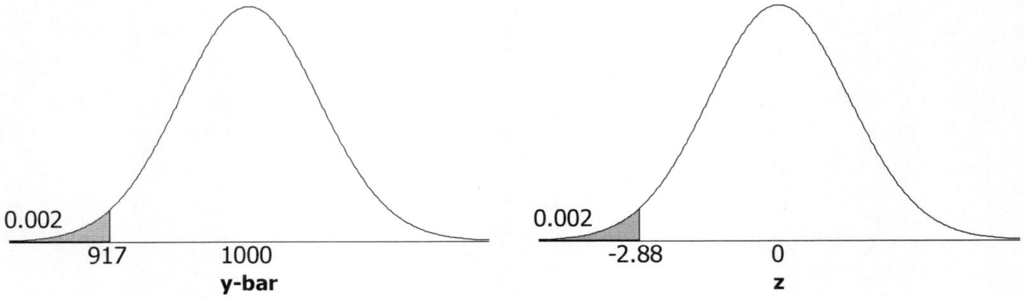

Figure 5.13: Distribution of the average weight of 12 lettuces (left panel)and standardised normal curve (right panel)

Note that, if the population from which the measurement arose is not normally distributed then you will need to be careful when calculating probabilities. If the sample size, n, is large enough then, by the Central Limit Theorem, the distribution of averages will be approximately normal. In this case your probability will be approximately correct. As we have seen in this chapter, "large enough" depends on the underlying distribution.

The distribution of a proportion

Often, a population proportion is the parameter of interest, rather than a population mean. For example, you might wish to know the proportion of undergraduates at a university who own PCs (as opposed to a MAC). This question could be answered by undertaking a sample survey of, say, 200 selected students from the population, and finding out how many said 'yes' to the question. If 132 said 'yes', your estimate of the population proportion would be 132/200, or 0.66.

You may also wish to know the likelihood of getting a sample proportion greater than (or less than) some specified value, given a known population proportion. For example, suppose you were told that 60% of the undergraduates at the university in question actually owned a PC. Given this information, what is the probability that a sample of 20 students has a majority owning PCs?

This type of question may be answered using the procedure described in the preceding section. The method is based on the fact that a proportion is really just a special case of an average, in which the measurements are coded as 0 or 1.

To see how this works, suppose that a student who owns a PC is given the value 1, and a student who does not own a PC is given the value 0. Thus if the data in a sample of size n are assigned the values $y_1, y_2, \ldots, y_n$, then y_i is 0 if member i does not own a PC, and y_i is 1 if member i does own a PC.

Using this notation, the proportion of PC owners in the sample (p, say) is given by the sample mean, that is

$$p = (y_1 + y_2 + \ldots. + y_n) / n \qquad (5.4)$$

This formula applies in general: the proportion of members of a sample or population having a particular attribute is equal to the mean of the sample or population, where the values are coded as 0 if the attribute is absent and 1 if the attribute is present.

Now let us denote the population proportion by π. If a sample of size n is taken, it may be shown that the sample proportion p has mean π and standard deviation

$$\text{st.dev}\ (p) = \sqrt{\frac{\pi(1-\pi)}{n}} \qquad (5.5)$$

This formula is similar to Equation 5.2, which gives the standard deviation of a sample mean. In fact Equation 5.5 is the special case of Equation 5.2 that arises when σ is $\sqrt{\{\pi(1-\pi)\}}$.

As Equation 5.4 indicates, p is the average of n components. According to the central limit theorem described earlier in this chapter, the distribution of p is thus approximately normal. Moreover, if p is standardised by subtracting its mean and dividing by its standard deviation, the distribution of the resulting z score is approximately standardised normal. This z score is thus defined as

$$z = \frac{p-\pi}{\sqrt{\pi(1-\pi)/n}} \qquad (5.6)$$

Now we can compute the probability that a majority of students in a class of 20 (assumed to be a random sample from the student population) own PCs. Given that the population proportion of PC owners is $\pi = 0.6$, the sample size is $n = 20$, and the sample proportion in question is $p = 0.5$, Equation 5.6 gives

$$z = \frac{p-\pi}{\sqrt{\pi(1-\pi)/n}} = \frac{0.5-0.6}{\sqrt{0.6(1-0.6)/20}} = -0.913\,.$$

As shown in Figure 5.14, the probability that p is greater than 0.5 is the same as the probability that z is greater than –0.913. To determine this proportion from a table of the standardised normal distribution, it is easiest to compute first the area under the curve below –0.913 (which is the same as the area under the curve above 0.913), and then to subtract this result from 1. If 0.913 is rounded to 0.91 to facilitate using the table, the required proportion is thus 1 − 0.181, that is, 0.819.

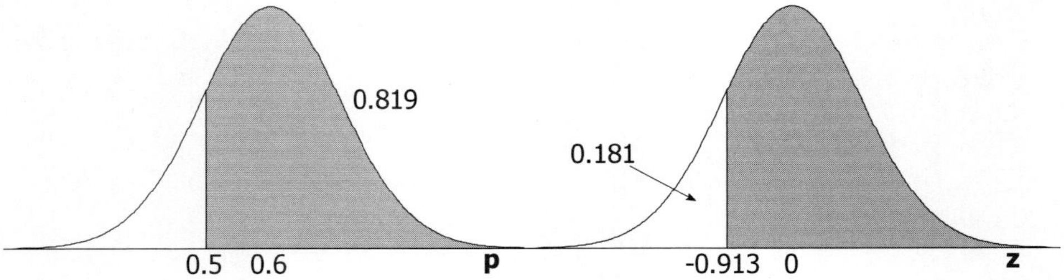

Figure 5.14: Distribution of the proportion of PC owners in a sample of 20

Further statistical applications

Many problems in Statistics involve estimating probabilities based on sample proportions. If the sample size is large enough, the normality approximation may be used, and the problem involves computing an appropriate area under the normal curve. Illustrations are given in this section.

Suppose you know for a fact that 30% of men in a particular population are smokers. What is the probability that a sample of 150 from this population contains at least 40 smokers?

This problem is very similar to the preceding one. Let p be the proportion of smokers in a sample of 150. The corresponding population proportion is $\pi = 0.3$. According to the theory, the distribution of p is approximately normal with mean 0.3 and standard deviation (using Equation 5.5) $\sqrt{(0.3\times0.7/150)} = 0.0374$.

Now, the number of smokers in the sample is 40 or more if the *proportion* of smokers in the sample, p is 40/150 = 0.2667 or more. Using Equation 5.6, the z score is

$$z = \frac{p-\pi}{\sqrt{\pi(1-\pi)/n}} = \frac{0.2667-0.3}{0.0374} = -0.890,$$

so the required probability is the area to the right of $z = -0.89$, as shown in Figure 5.15. Using Table A1, the area in the tail of the standardised normal distribution beyond $|z| = 0.89$ is 0.1867.

Since the distribution is symmetric, this is the same as the area in the left tail below $z = -0.89$. Since the total area under the curve is 1, then the area under the curve to the right of $z = -0.89$ is $1-0.1867 = 0.813$. This is the required probability.

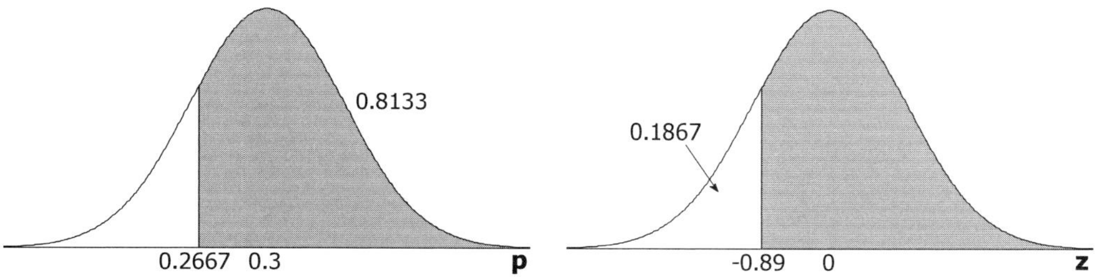

Figure 5.15: Distribution of the proportion of smokers in a sample of 150

As a further illustration, consider standby passengers hoping to get seats on an airline flight. The plane has 300 seats and is fully booked, with 35 would-be passengers on standby. However, it is known from experience that only 90% of passengers with bookings show up.

What is the probability that all 35 standby customers will get seats?

To solve this problem, let p denote the proportion of passengers with bookings who fail to show up. According to the theory, the distribution of p is approximately normal with mean 0.1 and standard deviation (using Equation 5.5) $\sqrt{(0.1\times0.9/300)} = 0.01732$.

All standby customers will get seats if p is at least 35/300 = 0.1167. Using Equation 5.6, the z score is

$$z = \frac{p-\pi}{\sqrt{\pi(1-\pi)/n}} = \frac{0.1167-0.1}{0.01732} = 0.964,$$

so the required probability is the area to the right of $z = 0.964$, as shown in Figure 5.16. Using Table A1, the area in the tail of the distribution beyond $|z|$ = 0.96 is 0.1685. This is the required probability. The probability that all standby customers will get seats is about 1 in 6.

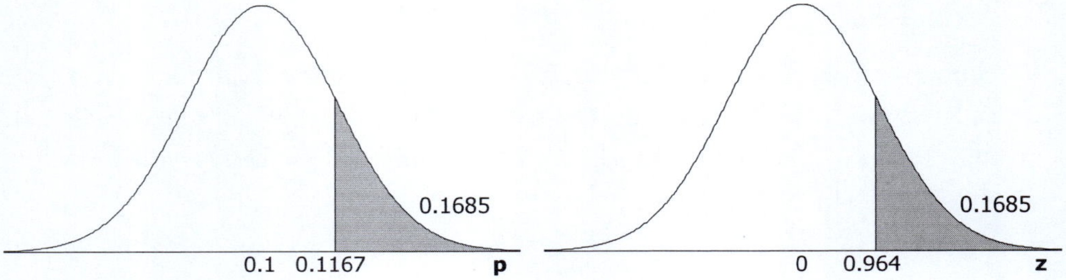

Figure 5.16: Distribution of the proportion of no-show airline passengers

It is important to note that the probabilities calculated above are only approximate if the sample size is large enough to apply the Central Limit Theorem. In the context of a proportion it is a reasonable approximation if both $n\pi$ and $n(1-\pi)$ are both at least 5. So in the examples involving PC ownership and male smokers the probabilities are reasonable.

Summary

This chapter is concerned with a fundamental law of Statistics, namely, that averages tend to be normally distributed. For example, you saw how the average of two or more independent random numbers, each uniformly distributed, has a distribution that is concentrated in the middle. And you saw how this theory applies to distributions of averages of naturally occurring quantities, such as the amount of sunshine in a day.

You also saw how the standard deviation of an average is less than the standard deviation of an individual measurement. As a result, averages provide more accurate determinations than individual measurements.

You may state this result more precisely by saying that the standard deviation of a sample average (called its *standard error*) is proportional to the reciprocal of the square root of the sample size.

Thus, to double the accuracy of a sample mean, the sample size needs to be not just doubled, but quadrupled.

If you may assume that the distribution of the data in a sample is normal, it follows that the sample mean also has a normal distribution. This fact may be used to compute the probability that a sample mean will exceed (or fail to exceed) a given value. The methods for computing probabilities described in the preceding chapter may then be applied.

Finally, you saw how the same methods are used to compute probabilities associated with sample proportions. These results are based on the fact that a proportion may be viewed as the special case of an average that arises when the data are restricted to taking two values. However, these results are only approximate, and depend for their validity on the sample size being sufficiently large.

Homework problems

1. Refer to Homework Problem 7 from Chapter 4, where a tennis club buys their ball from a manufacturer who designs their balls to have the following specifications (each being normally distributed):

   weight: mean = 57.4 grams, st.dev. = 0.4 grams;
   bounce: mean = 143 cm, st.dev. = 3 cm.

   a. A club member decides to test his can of three balls and finds that they have a mean weight of 58.0 g. What is the likelihood of getting such a high average weight for the balls in a can?

   b. The club receives a delivery of balls and is concerned that they may not have come from the usual manufacturer. A bouncing test is conducted on a sample of 60 balls and the average bounce is found to be 141 cm. What is the likelihood that the average is as low as this, assuming that the balls did come from the usual manufacturer?

   c. Given the result obtained in part b. what decision should the club make? Explain.

2. A university lecturer claims that he is plagued by bad luck, because most times he goes skiing the weather is rainy. A colleague in the Statistics Department is skeptical of this claim, stating that if the probability of rain on any given day during the ski season is as high as 25%, it would not be unusual to get 9 or more rainy days out of 17. Compute this likelihood using the normal distribution, and comment on the validity of the approximation.

3. It is claimed that the average time to travel from an inner suburb to the city during the peak hour is normally distributed with mean 20 minutes and standard deviation 2 minutes. The times taken by 16 (randomly selected) motorists to travel from this suburb to the city during the peak hour were recorded as follows (in minutes):

   21.6 24.2 18.6 19.7 21.7 26.3 20.5 22.4
   29.7 20.6 19.1 23.4 24.6 18.8 22.8 20.1

   a. What is the probability of getting a sample with a mean this high?

   b. Further investigation reveals that the time of 29.7 minutes is atypical because it occurred when there was a strike by public transport workers. Repeat (a) with this outlier removed.

4. Refer to Homework Problem 8 from Chapter 4, concerning the red striped tiger worms and their voracious appetite for kitchen scraps.

   a. One retailer claims that his can of worms can eat 1.1 kilograms of waste in a day. What is the likelihood of this?

   b. What is the probability that a can of worms can eat more than 6.5 kilograms of waste in a week, given that they do not take time off for weekends?

5. On 14 June 1993 *The Sydney Morning Herald* reported that a study had found that one in five children in local schools had repeated at least one year of primary school before the end of Year 6.

   a. In a school with 80 children completing Year 6, how many of these children would you expect to have repeated a year?

   b. In a local high school where 200 children enter Year 7, what is the probability that more than 50 have repeated a year in primary school?

6. A report by the Institute of Family Studies cited in *The Sydney Morning Herald*, 30 August 1993, stated that in 1991, 16.6% of families comprised a single parent with dependent children. In a town with 5000 families, what is the probability that more than 1000 of these families will comprise single parents with dependent children?

7. A University department wishes to set up a computer laboratory for its postgraduate students. There are 22 postgraduate students, each of whom uses a computer 50% of the time. How many computer terminals are needed to ensure that the chance that a student will not find an available terminal is less than 0.1?

8. To graduate in Cosmology from *Toughs* University, a student must complete a practical assignment within 2 ½ hours. The average time taken by individual students to complete this assignment is known from experience to be 2 hours with a standard deviation of 1 hour. But students are permitted to form teams containing up to four members. The rule for teams is that they must still tackle the assignment independently, and the whole team passes if the team's average time to complete the assignment is less than 2 ½ hours, but the whole team fails if this average exceeds 2 ½ hours.
   a. What is the probability that an individual who is not a member of a team will pass? (Assume a normal distribution for the individual completion times.)
   b. What is the chance that a team of four students will pass? Do you still need to assume that the completion times are normally distributed to answer this question? Explain.
   c. What is the chance that the four students will all pass if they do not form a team? Again state whether you need to assume a normal distribution for the completion times.
   d. Four students meet to discuss strategy before the test. Should they form a team? Why?
   e. The students think that they can improve their chances of getting through if they are allowed to form larger teams, and the University is willing to allow this, but insists that larger teams must achieve an average completion time of 2 ¼ hours. Is this reasonable? Explain.

9. Ray Dundee has a hydroponics farm on the Gold Coast where he grows strawberries using tissue culture. He is considering a deal with Royal Asian Airlines (RAA) to provide superior quality strawberries for their international flight catering. The fruit are to be shipped in punnets each containing exactly 12 strawberries.

   A survey of RAA passengers reveals that they prefer their strawberries to weigh 30 gm, and will complain if they are served strawberries weighing less than 22 gm. However using modern technology Ray can produce strawberries with a mean weight of 35 gm and a standard deviation of 5 gm, so he agrees to the deal.

   a. Assuming the weights of Ray's strawberries are normally distributed, what is the proportion weighing less than 22 gm?
   b. RAA monitors Ray's strawberry shipments using the old-fashioned quality control method of selecting a punnet at random from each shipment and rejecting the whole shipment if the punnet weighs less than 360 gm. What is the probability that a shipment will be rejected?
   c. A Japanese restaurateur whose customers want even larger strawberries approaches Ray. Ray agrees to sell 40% of his strawberry crop to this new client, these comprising all fruit above a specified weight. What is this weight?
   d. Ray now realises that he may be in trouble, because the mean weight of the strawberries he is now supplying to RAA has been reduced as a result of the deal with the Japanese restaurateur. He doesn't know how to compute the mean of the new distribution, but he knows how to compute its median. What is this median?
   e. Ray desperately needs to do a simulation to find out if he can meet the RAA quality inspection requirement, but he doesn't know how. His prayers are answered when his niece, a student who is studying Statistics at Macquarie University, arrives for a holiday on his farm. She installs a statistical package on his computer and uses it to generate, summarise, and display the weights of a random sample of 960 strawberries (80 punnets) to be supplied to RAA, obtaining the values 31.91gm for the mean and 3.13 gm for the standard deviation. Use these parameters to estimate the probability that a shipment will be rejected.
   f. Ray is providing a shipment to RAA every week, and he is very worried by this result. What should he do now?

CONFIDENCE INTERVALS

CHAPTER 6

6 Confidence Intervals

In this chapter we show how you can draw conclusions about a population, based on the results observed in a sample.

Whenever you do an empirical study you start with a sample. The sample contains all the data you have collected. You would like to make a conclusion. But this conclusion refers to the *population,* not the sample. You may be able to describe the sample precisely, but your statement about the population is subject to uncertainty.

A good way of describing the extent of uncertainty remaining in a population after taking a sample of data from it is to give a *confidence interval* for each population parameter of interest. A confidence interval is simply a range of values within which you are reasonably certain that an unknown population parameter is located. For example, suppose that a sample of 400 voters were asked for their opinion on whether Australia should change its flag. If 220 favoured a change, it may be possible to conclude with a high degree of certainty that the support in the electorate for this proposition is between 50% and 60%. The interval (50%, 60%) is thus a confidence interval for the unknown population percentage favouring the change.

This chapter describes how to compute and interpret confidence intervals. We illustrate the concepts by considering in detail the estimation of two population parameters of interest: a population proportion and a population mean.

First, some formulas need to be developed. These formulas are based on the theoretical result described in the preceding chapter, namely, that the distribution of the average of a set of independent measurements is, at least approximately, normally distributed with a specified standard deviation.

Confidence interval for a population proportion

Suppose that a large car insurance company has set up a database containing up-to-date information about its dealings with clients, including their claims. As a service to its clients and other interested parties, the company will supply, at a price, any information from its database, with the proviso that it cannot reveal names and other facts that may identify individuals personally.

Now suppose you would like to know the overall proportion of clients who made some claim during the last year. The company has this information, but upon request will only provide the proportion of claimants in a randomly drawn sample from its population of clients, and it charges an amount $\$n$ for a sample of size n.

This situation is not uncommon. In fact this is how you get detailed demographic information from the Bureau of Statistics. It costs money to gain information and the larger the sample the more it costs.

Suppose you are prepared to spend \$400 to get an estimate of the proportion of car insurers who made claims in the year. So the company duly selects 400 clients from its database and informs you that 94 clients in this sample made a claim during the previous 12 months. Your estimate of the proportion of claimants is thus 94/400 = 0.235.

How accurate is this estimate? How close is it to the proportion of claimants in the client population? Could the true proportion be as low as 0.2? Or could it be as high as 0.3?

To answer the question, you could take more samples and see how the sample proportion varies from sample to sample. If another sample of 400 gave 92 claimants (proportion 0.23) you might be inclined to believe that a sample of 400 is large enough to give a reasonably accurate estimate. But if another sample of 400 gave only 68 claimants (proportion 0.17) you would not feel so confident about estimating the true population proportion. You could take more samples, but remember that each additional sample costs more money.

Actually, there is a cheaper solution. It is based on the theory of distributions of averages described in the preceding chapter. Recall (Equation 5.6) that a sample proportion, based on a sample of size n from a population in which the corresponding proportion is π, is approximately normally distributed with mean π and standard deviation $\sqrt{\{\pi(1-\pi)/n\}}$. So using a property of the normal distribution, 95% of the time the sample proportion is within $1.96\times\sqrt{\{\pi(1-\pi)/n\}}$ of π. Putting it another way, the population proportion is probably (95% probable, to be precise) no further than $1.96\sqrt{\{\pi(1-\pi)/n\}}$ from the sample proportion p.

Consequently the unknown population proportion π is probably within the interval

$$\left(p-1.96\sqrt{\frac{\pi(1-\pi)}{n}},\ p+1.96\sqrt{\frac{\pi(1-\pi)}{n}} \right) \tag{6.1}$$

The only problem with Formula 6.1 is that it involves the unknown population proportion π . But if π was known, there would be no need to take a sample: the reason for taking the sample is to get an estimate for π!

The solution is actually quite simple. Just replace π by p in the formula. This will reduce the accuracy a little, but remember that Formula 6.1 is already an approximation because it is based on the assumption that the sample proportion has a normal distribution. The accuracy increases as the sample size n increases.

To summarise, given a sample proportion p based on a sample of size n, it is 95% probable that the population proportion π is within the interval

$$\left(p-1.96\sqrt{\frac{p(1-p)}{n}},\ p+1.96\sqrt{\frac{p(1-p)}{n}} \right) \tag{6.2}$$

This interval is called a *95% confidence interval* for the unknown population proportion.

It is important to note that Formula 6.2 is an approximation whose accuracy increases as n increases. In practice, it is a reasonable approximation if both $n \times p$ and $n \times (1-p)$ are at least 5.

Now you can answer the question posed at the beginning of this section. If a sample of 400 insurers indicates 94 claimants, the sample proportion is $p = 94/400 = 0.235$, the sample size is $n = 400$, and Formula 6.2 gives

$$\left(0.235-1.96\sqrt{\frac{0.235(1-0.235)}{400}},\ 0.235+1.96\sqrt{\frac{0.235(1-0.235)}{400}} \right)$$

This reduces to

$$(0.235-0.042,\ \ 0.235+0.042),$$

or (0.193, 0.277). So you can say, with 95% confidence, that π is between 0.193 and 0.277. This result is a valid approximation because both $n \times p = 94$ and $n \times (1-p) = 306$ are at least 5.

***Exercise 6.1**: Obtain a 95% confidence interval for the proportion of persons in Australia who would like to see their national flag changed, assuming that a random sample of 400 such persons gave 220 preferring the change.*

It is informative to make a graph of a confidence interval by denoting the estimate as a point and the confidence interval as a horizontal line. Figure 6.1 shows a graph of the 95% confidence interval for the proportion of insurance claimants in the population, based on the information that 94 claimants were observed in a sample of size 400. You can see easily from this graph that the probable range of the unknown population proportion is from a little below 0.2 to about 0.275.

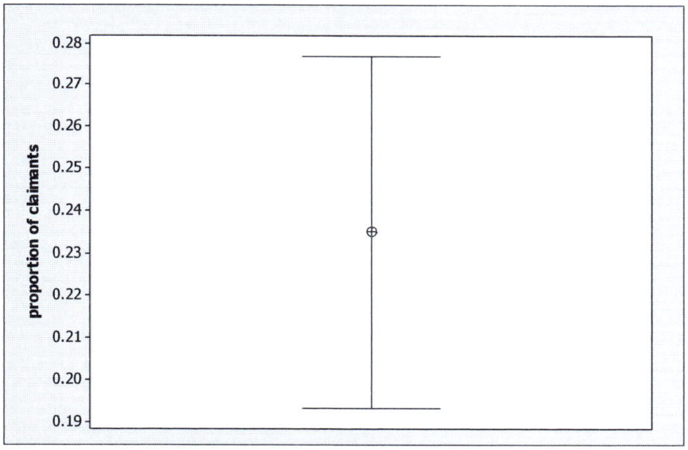

Figure 6.1: Graph of confidence interval for a population proportion

***Exercise 6.2**:*

A new poll, based on a representative sample of 1000 Australians, finds that 55% want to see the Union Jack removed from the Australian flag. Graph a 95% confidence interval for the proportion of persons in Australia who would like to see this flag modification.

According to Equation 5.5, the standard error of a proportion is inversely proportional to the square root of the sample size. This means that if the sample size is doubled, the length of the confidence interval is not halved, but only reduced by a factor $1/\sqrt{2}$ (approximately 0.71). To halve the length of the confidence interval (thus doubling the accuracy of the estimate) the sample size needs to be increased by a factor of 4.

Exercise 6.3*: Suppose you wish to estimate the proportion of car insurance claimants to within 1% (that is, you would like a 95% confidence interval of length 2%). How large a sample is needed? Solve this problem assuming that the unknown population proportion is (a) 0.3, (b) 0.2, and (c) 0.1.*

Summary: Confidence interval for a population proportion

To summarise, an approximate 95% confidence interval for an unknown population proportion is given by the interval

$$\left(p-1.96\sqrt{\frac{p(1-p)}{n}},\ p+1.96\sqrt{\frac{p(1-p)}{n}}\right),$$

where p is the sample proportion and n is the sample size.

This result is approximate because it is based on the assumption that the distribution of sample proportions is normal, and it uses the approximation $\sqrt{\{p(1-p)/n\}}$ rather than the exact result $\sqrt{\{\pi(1-\pi)/n\}}$ for the standard error of a proportion. The approximation improves as n increases, and, as a rule of thumb, is valid provided both $n \times \pi$ and $n(1-\pi)$ are at least 5.

Confidence interval for a population mean

We now have a method for estimating a population proportion based on a sample. Starting with the sample proportion p, this method gives an interval, called a confidence interval, which is centered at p and is likely (with probability close to 95%) to contain the population proportion.

In this section we consider the estimation of a population mean.

The example we started with in this chapter involves estimating the proportion of car insurers who made a claim during the preceding year. Continuing with this application, suppose you want to know what is the average premium paid by the insurers. Of course the insurance company has this information in its database, but let us suppose, as we did before, that the company will provide sample data only, and charges for this information.

So you could take a sample of n customer premiums from the company's database, and calculate the sample mean $\bar{y}$. What is needed now is a confidence interval for the population mean.

Let us recap the argument used in the preceding section to obtain a confidence interval for an unknown population proportion.

First, we use the fact that if samples of size n are taken repeatedly, the distribution of the corresponding sample proportions is approximately normal, with mean π (the population proportion), and standard deviation given by the formula

$$\text{st.dev}\,(p) = \sqrt{\frac{\pi(1-\pi)}{n}}$$

Next, the unknown population proportion π is replaced in this formula by p, the estimate based on the sample. Finally, we use the fact that a normal distribution has 95% of its area within 1.96 standard deviations of its mean. This means that the discrepancy between the sample proportion p and the population proportion π is probably less than $1.96\sqrt{\{p(1-p)/n\}}$. ('Probably' means 'with probability 0.95'.)

In other words, it is 95% probable that π lies between the values $p - 1.96\sqrt{\{p(1-p)/n\}}$ and $p + 1.96\sqrt{\{p(1-p)/n\}}$. However, this result is not exact because (a) the normal distribution is only an approximation to the distribution of p, and (b) it uses an estimate of the standard deviation.

A very similar argument may be used to get a confidence interval for an unknown population mean. For the moment, let us assume that the population generating the sample is itself normal.

First, we use the fact that if samples of size n are taken repeatedly, the distribution of the corresponding sample means is normal, with mean μ (the population mean), and standard deviation given by the formula

$$\text{st.dev}\,(\bar{y}) = \frac{\sigma}{\sqrt{n}}$$

As before, we use the fact that a normal distribution has 95% of its area within 1.96 standard deviations of its mean. Again it follows that the discrepancy between the sample mean $\bar{y}$ and the population mean μ is, with probability 0.95, less than $1.96\,\sigma/\sqrt{n}$. In other words, μ "probably" lies between the values $\bar{y} - 1.96\,\sigma/\sqrt{n}$ and $\bar{y} + 1.96\,\sigma/\sqrt{n}$.

This method gives an exact 95% confidence interval for μ, namely

$$\left(\bar{y} - 1.96\frac{\sigma}{\sqrt{n}},\ \bar{y} + 1.96\frac{\sigma}{\sqrt{n}}\right) \tag{6.3}$$

The only assumption made here is that the population generating the sample is normally distributed. Although, if the sample size n is large, then the Central Limit Theorem will apply and the distribution of sample means will be approximately normal.

As an illustration, suppose that, for a sample of 50 car insurers the average premium is \$1875, and that you are given that the standard deviation of premiums in the target population is \$450. Thus $\bar{y}$ is 1875, n is 50, and σ is 450. Substituting these values into Formula 6.3, the 95% confidence interval for the population mean μ is thus

$$\left(1875 - 1.96\frac{450}{\sqrt{50}},\ 1875 + 1.96\frac{450}{\sqrt{50}}\right),$$

which is (\$1750, \$2000).

There is a problem with Formula 6.3. As you can see, it involves the population standard deviation σ. Now, if the population mean μ is unknown, it is unlikely that the population standard deviation σ is known. So Formula 6.3 may not be very useful in practice.

Recall that the same sort of difficulty arises in the derivation of a confidence interval for a population proportion. In that case we simply replaced the unknown population parameter by its estimate based on the sample. By the same token, if σ is unknown, you can just replace it in Formula 6.3 by the corresponding sample standard deviation s, to give the formula

$$\left(\bar{y} - 1.96\frac{s}{\sqrt{n}},\ \bar{y} + 1.96\frac{s}{\sqrt{n}}\right) \tag{6.4}$$

Formula 6.4 gives an *approximately* valid 95% confidence interval. The approximation improves as n increases, but if n is small you would not expect the formula to be very accurate. How large should n be? If the sample size is 10, can Formula 6.4 be used?

Before answering this question, let us do a simulation study. Suppose we repeatedly generate samples of size 10 from a normal distribution with given mean and standard deviation. Suppose the population mean is $\mu = 150$ and the population standard deviation is $\sigma = 30$.

Figure 6.2 shows the confidence intervals for 20 samples using Formula 6.4, with the sample means and sample standard deviations listed. You can see that the sample means vary from a minimum of 136.8 to a maximum of 167.2, and that all but two of the samples give a confidence interval that includes the population mean (150). The lengths of the confidence intervals also differ substantially, depending as they do on the standard deviations of the samples.

$\bar{y}$	s
151.8	24.6
152.9	31.3
136.8	34.7
150.9	28.5
161.5	28.7
151.5	34.2
143.8	31.8
136.8	31.8
161	35.9
149.1	27.3
158.5	35.7
145.1	24.6
151	22.3
145.2	22.5
149.3	40.7
163.1	26.3
161.8	17.7
142.3	20.1
167.2	18.1
156	31.2

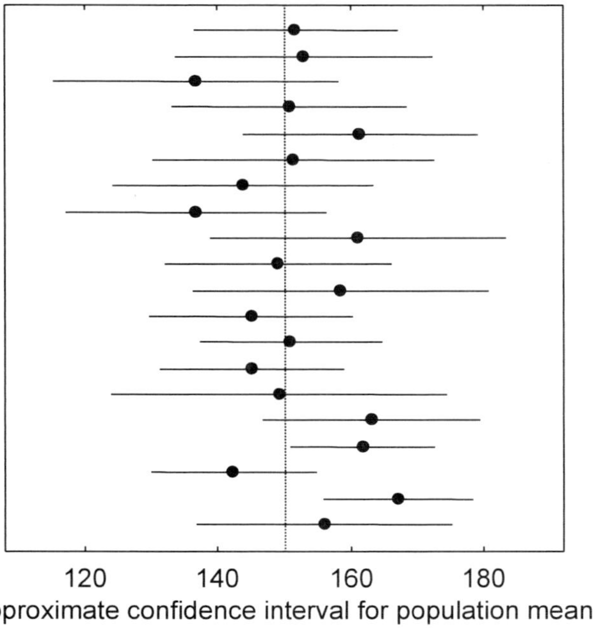

Figure 6.2: Confidence intervals from simulation study using Formula 6.4

The confidence intervals are not accurate because they are based on the sample standard deviations (which vary from sample to sample), instead of the fixed (but unknown) population standard deviation. However, it may be shown that an accurate 95% confidence interval for a population mean is obtained simply by stretching the confidence interval given by Formula 6.4, using the formula

$$\left(\bar{y} - t_{n-1}\frac{s}{\sqrt{n}},\ \bar{y} + t_{n-1}\frac{s}{\sqrt{n}}\right) \tag{6.5}$$

In Formula 6.5, t_{n-1} is a function of the sample size, called a *critical value*, which increases as n decreases. If the sample size is very large there is no loss of accuracy in using Formula 6.4, so its value is 1.96. Figure 6.3 gives values of t_{n-1} for various values of n. Note that the smallest value n can have is 2, because if n is 1 the sample standard deviation cannot be computed.

n	t_{n-1}	n	t_{n-1}	n	t_{n-1}	n	t_{n-1}
2	12.71	12	2.201	22	2.080	35	2.032
3	4.303	13	2.179	23	2.074	40	2.023
4	3.182	14	2.160	24	2.069	45	2.015
5	2.776	15	2.145	25	2.064	50	2.010
6	2.571	16	2.131	26	2.060	60	2.001
7	2.447	17	2.120	27	2.056	70	1.995
8	2.365	18	2.110	28	2.052	80	1.990
9	2.306	19	2.101	29	2.048	90	1.987
10	2.262	20	2.093	30	2.045	100	1.984
11	2.228	21	2.086	31	2.042	∞	1.960

Figure 6.3: Critical values for the 95% confidence interval for the mean of a normal population with unknown standard deviation

From Figure 6.3, if n is 10, the critical value is 2.262. Figure 6.4 shows a graph of the 95% confidence intervals for the same simulation study as we did before, using the exact formula 6.5 rather than the approximate formula 6.4. The lower and upper limits of the confidence intervals (denoted as 'min' and 'max') are also listed in Figure 6.4. Compared with the intervals in Figure 6.2, each confidence interval in Figure 6.4 is stretched by the factor 2.262/1.96 = 1.154. As a result, the fourth interval from the bottom now includes 150, and only one of the 20 confidence intervals does not include this value.

min	max
134.2	169.4
130.5	175.3
112	161.6
130.5	171.3
141	182
127	175.9
121.1	166.5
114	159.5
135.3	186.7
129.6	168.6
133	184.1
127.4	162.7
135.1	167
129.1	161.2
120.2	178.4
144.3	181.9
149.1	174.4
128	156.7
154.3	180.2
133.7	178.3

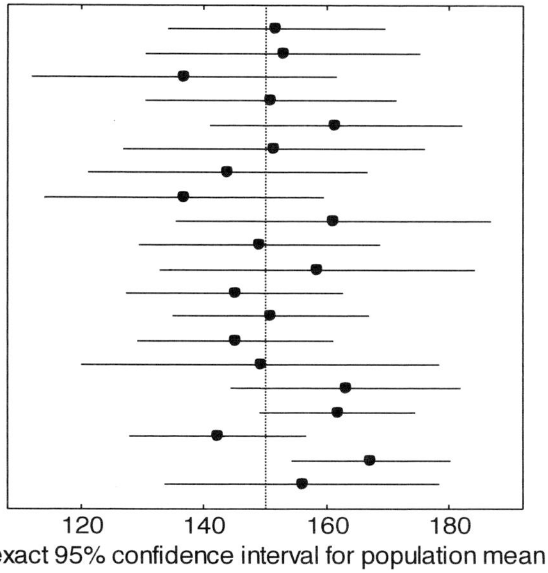

Figure 6.4: Confidence intervals from simulation study using Formula 6.4

Since the simulation study is based on 20 samples from a normal population having mean 150, you would expect, on average, 95% of the confidence intervals to include this value. So the result is as expected.

Of course if you were to repeat the simulation study you would not expect to find exactly 95% of the confidence intervals including the population mean. But you would expect the proportion of confidence intervals containing the population mean to be 0.95 *on average*.

Recall that the question raised in this section was 'how accurate is Formula 6.4 for a sample of size 10?' The answer is that it gives a confidence interval that is about 15% too narrow. In any case, you can get a 95% confidence interval that is exactly right, simply by replacing the value 1.96 in Formula 6.4 by the appropriate critical value taken from Figure 6.3.

As an illustration of the formula, let us obtain separate 95% confidence intervals for the body fat percentages for the men and women based on the data listed in Figure 2.1.

The four men's values (9.5, 7.8, 17.8 and 27.4) have mean $\bar{y}$=15.625 and standard deviation s = 8.984. The sample size is n = 4, so Figure 6.3 gives t_{n-1} = 3.182. Substituting these values into Formula 6.5, the 95% confidence interval for the population mean is

$$\left(15.625 - 3.182\frac{8.984}{\sqrt{4}}, 15.625 + 3.182\frac{8.984}{\sqrt{4}}\right),$$

which reduces to (1.33, 29.92).

For the 14 women in the sample, the mean $\bar{y}$ is 32.32 and the standard deviation is s = 4.90. The sample size is n = 14, so from Figure 6.3 you get t_{n-1} = 2.160. Substituting these values into Formula 6.5, the 95% confidence interval for the population mean is thus

$$\left(32.32 - 2.16\frac{4.90}{\sqrt{14}}, 32.32 + 2.16\frac{4.90}{\sqrt{14}}\right) = (29.49, 35.15).$$

The two confidence intervals are graphed together in Figure 6.5. You can see clearly from this graph that the mean body fat percentages for the male and female populations are quite different. In fact it is very likely that the men have a population mean below 30%, and that the women have a population mean above 30%.

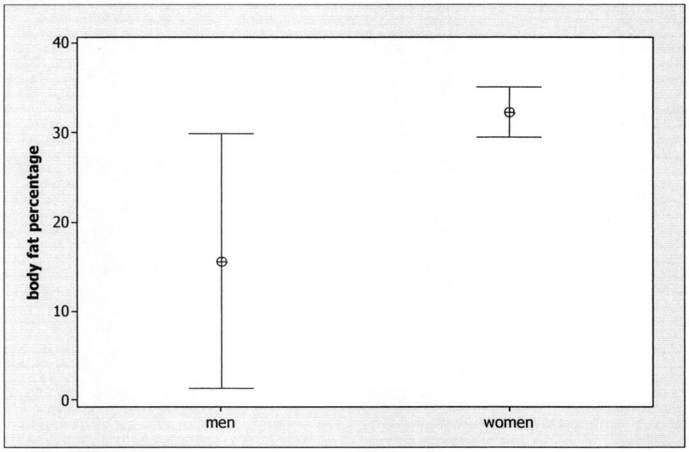

Figure 6.5: 95% confidence intervals for body fat percentages

The confidence interval for the mean of the male population is much wider than the corresponding confidence interval for the female population. There are two reasons for this:

a. the sample standard deviation for the fat percentages of the males (s = 8.98) is almost double the corresponding value for the females (s = 4.90);

b. the sample size for the males (n = 4) is much smaller than that for the females (n = 14).

The width of a confidence interval for a population mean is proportional to $s/\sqrt{n}$, so the combined effect of a. and b. is to make the confidence interval for the male population mean much wider than that for the females.

Exercise 6.4: *Obtain separate 95% confidence intervals for the mean ages of day and evening students, based on the data given at the beginning of the section entitled 'Measures of spread' in Chapter 3. These data comprise two samples each of size 10, each with sample mean 27, and with standard deviations 9.01 and 2.05, respectively.*

Normality Assumption

Note that formulas 6.3 and 6.5 require that the data come from a normal population. If the data are skewed, consider transforming them to improve the normal approximation before computing a confidence interval. A confidence interval for the mean of the original population may then be obtained by reversing the transformation.

Sample size calculation

One of the decisions you need to make when you do a study is to choose the sample size. If the sample is too small, the result may not be sufficiently accurate. On the other hand, the sample might be unnecessarily large, and is thus wasteful of resources.

The formulas given in this chapter may be used to calculate the sample size needed to obtain a specified accuracy. The accuracy depends on the width of the confidence interval: the shorter the confidence interval, the greater the accuracy. To be definite, we define the *accuracy* of estimation as **half** the width of a 95% confidence interval. Therefore, if an accuracy of 0.01 (or 1%) is required when determining a population mean, the width of the 95% confidence interval is 0.02 (or 2%).

Determining sample size for a population proportion

First consider estimating a population proportion. According to Formula 6.1, half the width of a 95% confidence interval for the population proportion is $1.96\ \sqrt{\{\pi(1-\pi)/n\}}$. If we use the symbol δ to denote the accuracy of estimation, we have the formula

$$\delta = 1.96\sqrt{\frac{\pi(1-\pi)}{n}} \qquad (6.6)$$

To determine the sample size needed to get a specified accuracy, Equation 6.6 needs to be expressed in terms of the sample size, that is,

$$n = (1.96/\delta)^2\pi(1-\pi) \qquad (6.7)$$

In practice, the population proportion π is unknown, so you need to assume a value for it. If π is completely unknown, giving it the value 0.5 in Equation 6.7 gives an upper bound for the sample size, because $\pi(1-\pi)$ has its maximum value, 0.25, when π is 0.5.

For example, suppose the Traffic Department wishes to do a survey to find out, to an accuracy of 2%, the proportion of motorists exceeding the speed limit at a given highway location. How many motor vehicles do they need to clock?

In this case δ is 0.02. Assuming that nothing is known about the population proportion, take $\pi = 0.5$. Equation 6.7 gives

$$n = (1.96/0.02)^2\ 0.5(1-0.5) = 2401.$$

So a sample size of 2400 is needed.

This is a quite a large sample size. But it is a fact that very large sample sizes are generally needed to gain accurate estimates of population proportions.

If you have some prior knowledge about the value of the population proportion, the sample size may be reduced to some extent. For example, if it is known that only a minority (at most 25%) of motorists exceed the speed limit. Then substituting $\pi = 0.25$ into Equation 6.7, we obtain

$$n = (1.96/0.02)^2\ 0.25(1-0.25) = 1801.$$

The sample size needed is still quite large, so it seems that the result doesn't depend to a great extent on the value of π.

However, the required sample size depends very much on the value of δ. If an accuracy of 5% is sufficient (again assuming that nothing is known about π), Equation 6.7 gives

$n = (1.96/0.05)^2\ 0.5(1-0.5) = 384.$

Though still quite large, this is quite a lot smaller than the value 2401 obtained when δ is 0.02.

Exercise 6.5:

An opinion pollster needs to determine the proportion of voters saying that the Prime Minister is doing a good job to an accuracy of 3%. How many voters need to be sampled?

Exercise 6.6:

The Department of Health has set a requirement that at least 95% of infants should get basic immunisation treatment within 12 months of birth. It wants to do a survey to determine this percentage to within 1%. How many subjects should be surveyed?

Determining sample size for a population mean

According to Equation 6.3, half the width of the 95% confidence interval for a the mean of a normally distributed population is

$$\delta = 1.96\frac{\sigma}{\sqrt{n}}. \qquad (6.8)$$

This formula is based on an assumption that σ is known, so if this is not the case, you will need to assume some value for σ. Expressing the result in terms of n, Equation 6.8 may be re-expressed as

$$n = \left(1.96\frac{\sigma}{\delta}\right)^2. \qquad (6.9)$$

As an illustration, suppose that the Traffic Authority wishes to estimate the average speed of motorists on a particular stretch of highway to within 2 km/hour. Since the standard deviation of motorists' speeds is unknown, assume the value $\sigma = 5$. (This would be reasonable if speeds were distributed with a range of 20 km/hour, using the rule of thumb that the standard deviation of a distribution is approximately one quarter of the range.) Substituting $\delta = 2$ and $\sigma = 5$ into Equation 6.9, we obtain

$$n = (1.96 \times 5/2)^2 = 24.1.$$

So the Traffic Authority only needs to clock 24 motorists.

In general, much smaller sample sizes are needed to determine population means than are needed for population proportions. This is due to the fact that the distribution of a continuous variable generally contains more information than the distribution of a binary variable.

Exercise 6.7:

The Health Department wants to know the mean age at which infants get immunised, to an accuracy of one month. Assuming that these immunisation times are normally distributed with standard deviation 4 months, how large a sample is needed?

Summary

In this chapter you have learned how to compute a confidence interval for a population proportion and mean. A confidence interval is an interval within which you can say, with specified certainty, that the unknown population parameter is located. By convention, this degree of certainty is taken to be 0.95 or 95%, in which case the confidence interval is called a 95% confidence interval.

A 95% confidence interval may be computed from a sample. If you wish to estimate an unknown population proportion, an approximate 95% confidence interval is given by the formula

$$\left(p - 1.96\sqrt{\frac{p(1-p)}{n}},\ p + 1.96\sqrt{\frac{p(1-p)}{n}}\right),$$

where p is the sample proportion and n is the sample size. This approximation improves as n increases, and is not valid unless both $n \times p$ and $n\,(1-p)$ are at least 5.

If the population standard deviation σ is known, a 95% confidence interval for an unknown population mean is given by the formula

$$\left(\bar{y} - 1.96\frac{\sigma}{\sqrt{n}},\ \bar{y} + 1.96\frac{\sigma}{\sqrt{n}}\right),$$

where $\bar{y}$ is the sample mean and n is the sample size, as before. This formula is exact, but is based on an assumption that the population distribution is normal.

Since the population standard deviation is usually not known, a more useful formula for a 95% confidence interval for a population mean from a normal population is

$$\left(\bar{y} - t_{n-1}\frac{s}{\sqrt{n}},\ \bar{y} + t_{n-1}\frac{s}{\sqrt{n}}\right),$$

where s is the sample standard deviation, and t_{n-1} is a function of the sample size, called a *critical value*, which decreases as n increases (as shown in Figure 6.3). When the sample size is very large, t_{n-1} takes the value 1.96, so that the formula reduces to the one used when σ is known.

In general, the width of a confidence interval decreases in direct proportion to the square root of the sample size. This means that if the sample size is doubled, the width of the confidence interval decreases by the factor $1/\sqrt{2} = 0.71$. And if the sample size is quadrupled, the width of the confidence interval is halved.

You also saw how to calculate the sample size needed to estimate a population proportion and a population mean with a specified degree of accuracy. The accuracy of estimation may be defined as half the width of the 95% confidence interval. In general, the accurate determination of a population proportion requires a much greater sample size than a corresponding sample mean.

Homework problems

1. The Head of the Taxation Office wishes to know the proportion of higher income earners (persons with before-tax incomes of more than $100,000) with private medical insurance. A random sample of 500 higher income earners reveals that 120 have private medical insurance. Display a 95% confidence interval for the proportion as a graph.

2. During the first year of operation of a new golf course, professional golfers at three-par greens recorded four holes-in-one. It is estimated that during the same period, professional golfers teeing off on these greens made 41,000 shots. Use these data to estimate the probability that a professional golfer will strike a hole-in-one at a three-par green.

   a. Can you give a 95% confidence interval for this probability? If so, do so. Otherwise explain why you can't.

   b. At another green, two professional golfers strike holes-in-one with successive shots. This event is reported on a news program, with the statement that the probability of such a coincidence is 100 million to one. Do you think this is a reasonable estimate?

3. Assume that IQ scores are normally distributed in the population with standard deviation 14. A class of fourteen nine-year-olds has the following IQ scores:

   115 123 95 108 143 105 87 96 111 130 120 108 98 121

   a. Graph a 95% confidence interval for the mean of this population.

   b. Is the normality assumption reasonable?

4. In a study reported by Sternberg et al (1982), samples of cerebrospinal fluid were taken from 10 patients hospitalised for psychotic schizophrenia. The following levels of dopamine *b*-hydroxylase (DBH) activity, in nmol/(ml)(h)/(mg), were measured:

   0.0150 0.0204 0.0208 0.0222 0.0226
   0.0245 0.0270 0.0275 0.0306 0.0320

   Use these data to obtain a 95% confidence interval for the mean DBH level for hospitalised psychotic schizophrenics.

5. The ages of the subjects whose body fat percentages are given in Figure 2.1 (in the same order as the data in Figure 2.1) are:

   23 23 27 27 39 41 45 49 50 53 53 54 56 57 58 58 60 61

   Make a graph of a 95% confidence interval of the mean age.

6. Refer to Homework Problem 4 above. A further 15 schizophrenia patients were judged to be non-psychotic, and their DBH levels were

   0.0104 0.0105 0.0112 0.0116 0.0130 0.0145 0.0154 0.0156 0.0170 0.0180 0.0200 0.0200 0.0210 0.0230 0.0252

   Compute a 95% confidence interval for the mean DBH level in this population, and compare to the psychotic patients.

7. Use the data in Appendix B to obtain 95% confidence intervals for the proportion of women, born outside Australia, whose favourite drink is a non-alcoholic drink.

8. The tonnages of 25 British merchant ships in 1907 was reported by Floud (1973, Table 4.1) as follows:

   44 144 150 236 739 970 2371 309 679 26 1272 3246 1904 357 1080 1027 45 62 28 2507 138 502 1501 2750 192

   a. Compute a 95% confidence interval for the mean tonnage of British merchant ships in 1907.
   b. Check that the normality assumption is reasonable. If not, repeat (a) after taking logarithms of the data, and by converting back to tonnages by taking antilogarithms, obtain a 95% confidence interval for the geometric mean tonnage of the population.

9. The Minister for Health wishes to survey the country to determine the current percentage of their incomes privately insured families spend on medical insurance. A previous survey showed that these families spent 1.9% of their income, with a standard deviation of 0.4%. How large a sample is needed to determine the current percentage to within 0.1%?

10. A television company, with approximately 10% of the viewing audience, wishes to undertake a telephone survey to estimate the percentage of the population who prefers its channel. How many responses are needed to estimate this proportion to within 1%?

HYPOTHESIS TESTING

CHAPTER 7

7 Hypothesis Testing

In the preceding chapter you saw how to get a confidence interval for a population parameter, using data from a sample. A confidence interval is a good way of showing how accurately you can determine a population characteristic when all you have is one representative sample.

This chapter focuses on *p-values*. To get a p-value, you first need to make a *hypothesis* about a population parameter. For example, using known information relating to IQ (intelligent quotient) scores, you might make a hypothesis that the mean IQ score in the general population of Australians is 100. After taking a sample of Australians, you are in a better position to know whether this hypothesis is consistent with current data. A p-value is a measure of this consistency.

A p-value supplements a confidence interval as an additional measure of the uncertainty due to sampling variation.

P-values for testing hypotheses are illustrated using the same parameters as the ones considered in the preceding chapter, namely, population means and proportions. The concepts are developed using a population mean as the parameter of interest with known standard deviation and then applied to examples involving a population mean with unknown standard deviation and a population proportion.

Null hypothesis

Suppose you are told that IQ scores were designed to be normally distributed with a mean of 100 and standard deviation of 15. If there has been no change in IQ scores over the years, it would be reasonable to hypothesise that the mean IQ score since first described early in the 1900s has remained at 100.

This hypothesis is called a *null hypothesis* because it is a statement that there has been no effective change in the population parameter. The word *null* means 'no effect' and comes from the Latin word *nullus* for 'none'.

The subscript 0 is used to designate the value of a population parameter specified in the null hypothesis, called the *null value.* In this case the population parameter is the mean μ so the null value, μ_0, is 100. The null hypothesis is abbreviated as H_0. So the statement that the value of μ under the null hypothesis is 100 may be written as

$$H_0: \mu = 100$$

To test this hypothesis, suppose you take a sample of 50 Australians, measure their IQ scores and determine the mean IQ score for the 50 people in the sample. Finding 102, you compute a 95% confidence interval, which ranges from 97.84 to 106.16.

Since this confidence interval contains the value of 100, you could say that the sample is *consistent* with the null hypothesis, or that 100 is a *plausible* value for the population mean.

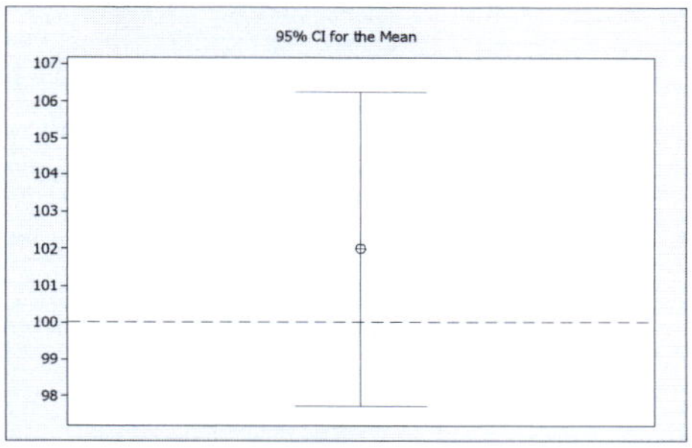

Figure 7.1: 95% confidence interval for a population mean, with hypothesised value

Now suppose that another sample of 50 Australians produces a mean of 105. The 95% confidence interval based on this new estimate is 100.84 to 109.16. In this case the confidence interval does not contain the null value, so you would be inclined to say that 100 is not a plausible value for the population mean.

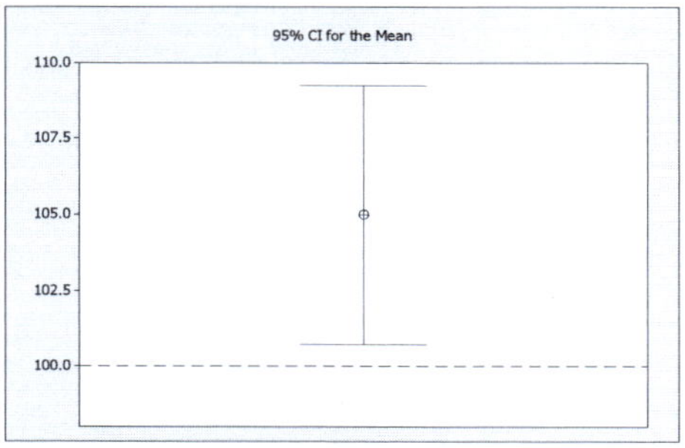

Figure 7.2: Alternative 95% confidence interval for a population mean, with hypothesised value

In general, a null hypothesis is a statement that a population parameter has a specified value. This specified value is chosen on the basis of studies that have investigated the question of interest. For a population mean, the null hypothesis may be expressed in general terms as

$$H_0:\ \mu = \mu_0, \qquad (7.1)$$

where μ_0 is the null value of the population mean (100 in the above example).

It is important that the null value should be specified in advance, before collecting the data. The null value should not be influenced by the data, because the data are used to test the null hypothesis.

P-value

In the preceding section we introduced the concept of a null hypothesis as a statement that a population parameter takes a specified null value, and we showed how its plausibility could be assessed graphically by comparing the null value with a confidence interval for the parameter.

In this section the concept of plausibility is made more precise, using a quantity called a *p-value.* A common way to get a p-value is to first define a more basic quantity called a *z-statistic.*

Z-statistic

A p-value is always associated with a null hypothesis and a sample. It is a measure of the extent to which the sample is consistent with H_0.

Consider the results graphed in Figures 7.1 and 7.2. In each case the null hypothesis states that the population parameter μ is 100, and a sample of 50 observations is taken from this population.

In the first sample the observed mean is $\bar{y}_1 = 102$ and the confidence interval based on this estimate includes 100, so the sample is consistent with the null hypothesis.

In the second sample the observed mean is $\bar{y}_2 = 105$ and the confidence interval based on this estimate does not include 100, so the sample is not consistent with the null hypothesis.

So a crude way of measuring the consistency, or agreement, between the sample and the null hypothesis is to say whether or not the confidence interval contains the null value.

A more informative measure of the extent of disagreement between the sample value and null hypothesis is the difference between the estimate and the null value. In the case of a population mean, this difference is given by $\bar{y} - \mu_0$.

The extent of the disagreement also depends on the length of the confidence interval, because if the confidence interval covers a wide range, it could contain the null value even when the difference between the estimate and the null value is quite large.

So a better measurement of the disagreement between the sample and the null hypothesis is obtained by *scaling* the difference. This could be done by dividing the difference by the length of the confidence interval or by the standard error. (As you saw in the preceding chapter, a 95% confidence interval has its length equal to 3.92 times the standard error, based on the normal approximation.)

Dividing the difference between the sample estimate and the null value by the standard error gives a quantity that is called a *z-statistic.* A formula for the standard error of a mean is given in a preceding chapter (Equation 5.2), namely

$$\text{st.dev.}(\bar{y}) = \frac{\sigma}{\sqrt{n}}$$

Using this formula, a *z*-statistic for comparing a sample mean with a hypothesised value is given by

$$z = \frac{\bar{y} - \mu_0}{\sigma / \sqrt{n}} \qquad (7.2)$$

Consider again the two hypothetical samples of Australians. In the first sample, $\bar{y}$ is 102. Since μ_0 is 100 and n is 50, Equation 7.2 gives

$$z = \frac{102 - 100}{15 / \sqrt{50}} = \frac{2}{2.1213} = 0.94$$

In the second sample $\bar{y}$ is 105, and the values of μ_0 and n are unchanged at 100 and 50, so the *z*-statistic now is (105–100)/2.1213, which is 2.36.

When the null hypothesis is true, it may be shown that the distribution of the z-statistic is approximately normal with mean 0 and standard deviation 1. So the next step in determining a p-value (at least approximately) is to use the definition of this standardised normal distribution.

Computation of p-value from z-statistic

In our examples, the p-value is defined in terms of the area in the two tails of a standardised normal distribution, as shown in Figure 7.3. The area in the left tail is the probability below $-z$ (assuming z is positive), and the area in the right tail is the probability above z, and these probabilities are the same because of the symmetric property of the normal distribution. The combined area is the probability of getting a value greater than z *in magnitude*, and this is the p-value.

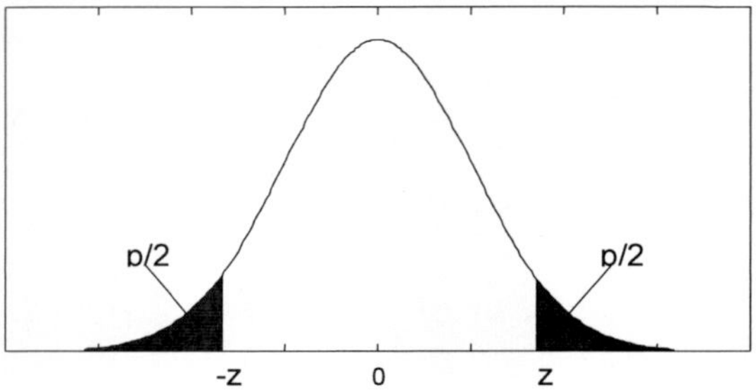

Figure 7.3: P-value based on the standardised normal distribution

Given the value of z, the corresponding p-value may be calculated from a table of areas under the normal curve, such as Table A1 shown in Appendix A.

For example, if z is 0.94 (as in the first sample of Australians), the area in the tail is found from Table A1 to be 0.1736. (Note that when z is negative, its absolute value is taken to get the area in the right tail.) Since the p-value is the area in both tails, this area needs to be doubled, so that the p-value is then $2 \times 0.1736 = 0.3472$.

Similarly, if z is 2.36 (as in the second sample of Australians), the area in the tail is 0.0091. So the p-value is this case is $2 \times 0.0091 = 0.0182$.

Note that the sign of the z-statistic does not affect its p-value. Suppose, for example, a third sample of 50 Australians is found to have a mean IQ of 98. In this case the z-statistic is -0.94, the tail area is 0.1736, and the p-value is 0.3472.

The p-value is thus the probability of getting a value under the standardised normal curve ***more extreme*** than the calculated z-statistic ***in either direction***. This is called a two-tailed test.

To summarise, for a two-tailed test, a p-value is obtained from a z-statistic by determining the area under the standardised normal curve in the two tails above $|z|$ and below $-|z|$. This area may be calculated from a table of normal curve single tail areas (such as Table A1).

Exercise 7.1:

A sample of 30 people produces a mean IQ of 106. Obtain a p-value for testing the hypothesis that the population mean is 100.

P-values and confidence intervals

Recall that a z-statistic is a measure of the discrepancy between an estimate of a population parameter based on a sample and its null value specified by a null hypothesis. The larger the z-statistic, the greater this discrepancy.

The larger the magnitude of the z-statistic, the smaller the p-value. You can see this from Table A1. The smallest value a z-statistic can have is 0, corresponding to a p-value of 1. The largest value given in the table for the z-statistic is 4, corresponding to a p-value of $2 \times 0.000032 = 0.000064$. If the magnitude of the z-statistic is greater than 4, the p-value will be even smaller than this.

How do p-values vary? To get some idea, consider some more examples.

Look again at the two samples of Australians' IQ scores graphed in Figures 7.1 and 7.2. In the first case the p-value is relatively large and the sample is consistent with the null hypothesis, while in the second case the p-value is relatively small and the sample data are not consistent with the null hypothesis.

Consider some more samples. Figure 7.4 shows 95% confidence intervals based on 20 more samples of 50 Australians, where the mean IQ scores (obtained by computer simulation) are as follows:

106.3 105.3 102.8 106.2 104.6 106.4 104.7 105.2 101.0 101.5

106.4 104.1 103.7 107.8 101.8 105.3 106.0 107.9 106.2 106.8

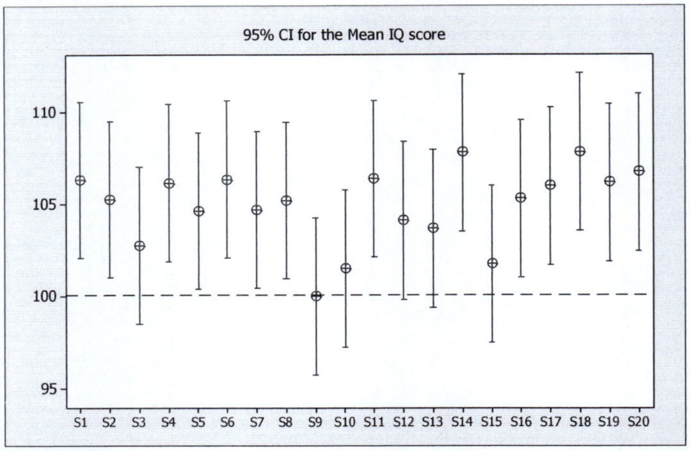

Sample	S1	S2	S3	S4	S5	S6	S7	S8	S9	S10
z-statistic	2.99	2.49	1.31	2,91	2.19	3,00	2.23	2.46	0.02	0.72
p-value	0.003	0.013	0.189	0.004	0.029	0.003	0.026	0.014	0.986	0.474

Sample	S11	S12	S13	S14	S15	S16	S17	S18	S19	S20
z-statistic	3.02	1.95	1.74	3.68	0.85	2.51	2.83	3.71	2.92	3.19
p-value	0.003	0.052	0.081	0.000	0.397	0.012	0.005	0.000	0.004	0.001

Figure 7.4: p-values for 20 samples of means with sample size 50 and null value 100

The null value is taken to be 100, and is denoted by the dotted horizontal line.

The p-values range from 0.000 to 0.986. The first sample (denoted by the leftmost confidence interval) has its confidence interval above the dotted line, and the corresponding p-value is 0.003. Most of the confidence intervals are entirely above the dotted line, with corresponding p-values smaller than 0.05. For six of the 20 samples the confidence intervals contain the null value 100, and these p-values are all greater than 0.05.

As a rough rule of thumb, a p-value is less than 0.05 if the corresponding 95% confidence interval for the population parameter does not contain the null value, and conversely, a p-value is greater than 0.05 if the 95% confidence interval contains the null value. This rule of thumb is exact for a mean since the standard error used to compute the confidence interval is based on the null value.

In Figure 7.4 most of the samples are in disagreement with the null hypothesis, and consequently most of the p-values are smaller than 0.05.

If the null value were the true value of the parameter in the population, you would not expect to get a small p-value.

Figure 7.5 shows the same confidence intervals as are graphed in Figure 7.4, but in this case the null value is taken to be 105, which happens to be the true value of the population mean used for the computer simulation. It was found that the Australians were in fact all University students whose mean IQ is higher than 100.

In this case the p-values are larger, ranging from 0.019 to 0.917. Only one confidence interval does not include the null value, and for this sample the p-value is 0.019.

When the null value is the true population parameter, the p-values tend to be evenly, or *uniformly,* distributed between 0 and 1 and consequently 5% of the p-values on average are less than 0.05. Figure 7.5 is in exact agreement with this rule: out of 20 samples taken from a population in which the mean is 105, just one has a p-value smaller than 0.05. This is what we would expect. However, this will not always be the case. We were just lucky in this sample of 20 samples to obtain the expected number of samples where the confidence interval does not contain the null value.

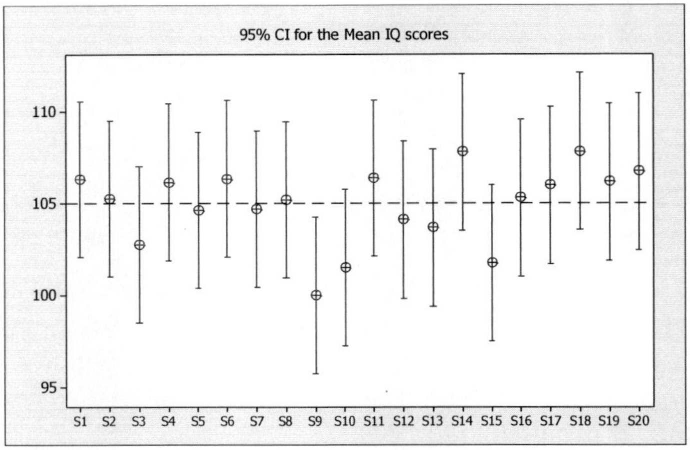

Sample	S1	S2	S3	S4	S5	S6	S7	S8	S9	S10
z-statistic	0.63	0.14	-1.04	0.55	-0.17	0.64	-0.13	0.10	-2.34	-1.64
p-value	0.526	0.892	0.296	0.580	0.865	0.521	0.899	0.917	0.019	0.101

Sample	S11	S12	S13	S14	S15	S16	S17	S18	S19	S20
z-statistic	0.66	-0.41	-0.61	1.33	-1.51	0.15	0.48	1.35	0.56	0.83
p-value	0.508	0.681	0.540	0.185	0.131	0.879	0.633	0.177	0.575	0.408

Figure 7.5: p-values for 20 samples of means with sample size 50 and null value 105

To summarise, if a null value specified by a null hypothesis is different from the true population parameter, you expect a sample to give a small p-value. Conversely, if the null value is equal to the true population parameter, you do not expect to get a small p-value.

The meaning of a p-value

A p-value may be interpreted as a likelihood, or probability. In fact the term p-value is an abbreviation for *probability* value. This probability is defined in terms of taking another sample, with the proviso that, for this sample, the null hypothesis is assumed to hold.

So suppose another sample is taken from a population in which the parameter of interest is equal to the null value. Then the p-value is simply the probability that this sample will give an estimate that is at least as extreme as the one you obtained the first time.

The p-value measures how rare (or unlikely) your sample is, assuming that the null hypothesis is true.

So if a sample gives a small p-value, there are two possible explanations:

a. The null hypothesis is true, but you obtained an atypical or unlikely sample.

b. Your sample is typical, but the null hypothesis is false.

Faced with a small p-value, the conventional approach adopted by scientific investigators is to choose b. because this approach *usually* leads to a correct decision. It does not always lead to the correct decision, because occasionally an atypical sample will arise.

The likelihood of making an incorrect decision depends on how small the p-value is. If the p-value is 0.1 and you decide to reject the null hypothesis, the probability of making a mistake is 0.1. But if you reject the null hypothesis based on a p-value of 0.01, the probability of making a mistake is only 0.01.

It is conventional to reject a null hypothesis when the sample data gives a p-value less than 0.05.

Consider the first sample graphed in Figure 7.4. The population mean assumed in the null hypothesis is $\mu = 100$. The sample mean, 106.3, differs from this assumed population mean by 6.3, and the p-value is 0.003. This is the probability that if another sample were taken from a population with mean 100, the difference would be *at least* 6.3. In other words, the probability is 0.003 that another sample of 50 from this population will give a mean *either* 106.3 or more, *or* 93.7 or less.

Now look at the third sample graphed in Figure 7.4. In this case the sample mean obtained is 102.8, and comparing this with the null value 100, a p-value of 0.189 is obtained. This is the probability that if μ were 100 another sample would give a difference of at least 2.8 from 100. So the probability is 0.189 that, in another sample of 50 from a population with mean 100, the observed mean will be either greater than or equal to 102.8 or less than or equal to 97.2.

So the first sample would lead to a conclusion that the population mean *is not* 100, since the likelihood of getting such an extreme sample is only

0.003 (less than 0.05). In the third sample however, it is more reasonable to conclude that the population mean *could be* 100, since the probability of getting such an extreme sample from a population with mean 100 is 0.189 (greater than 0.05).

Now the p-value is the likelihood that another sample will give a result at least as extreme as the one observed, assuming the null hypothesis is true. If this likelihood is small (by convention, smaller than 0.05) it is reasonable to *reject* the null hypothesis. But if the p-value is not small (that is, greater than 0.05), it is reasonable to conclude that H_0 *could be* true.

A p-value may be used to make a decision as follows:

c. If a p-value is small (conventionally less than 0.05) you *reject* the null hypothesis.

d. If a p-value is not small (0.05 or more) you say the null hypothesis *could be* true, ***but you do not accept it.***

So a p-value in itself does not give sufficient information to accept a null hypothesis. The sample size may be too small to give a conclusive result.

Alternative hypothesis

An alternative hypothesis also needs to be specified when conducting a hypothesis test. In each of the examples in the previous sections we were comparing our sample value with a hypothesised value with no indication, before the data were collected, of a direction for our alternative hypothesis. Therefore, in this situation, our alternative hypothesis will be:

$$H_1:\ \mu \neq \mu_0 \qquad (7.3)$$

However, consider a different situation whereby there is a claim that indicates a direction for the alternative hypothesis. For example, in the example involving the University students we might hypothesise that the mean IQ for this population is no different to that of the general population. Clearly, it seems reasonable to assume that University students may have an average IQ score that is higher than that in the general population. In this case our alternative hypothesis would be

$$H_1:\ \mu > 100$$

indicating that the resulting hypothesis test allows only for a rejection in a particular direction, in this case "greater than". Therefore the p-value will be on only one side of the distribution ie. the upper tail of the distribution in this case. We refer to this as a one-tailed test rather than a two-tailed test which were shown in previous examples when we calculated the p-values in both tails of the distribution.

To summarise, if there is a direction indicated in the research question then we perform a one-tailed test, whereas if there is no direction indicated then a two-tailed test should be performed. So, under a null hypothesis of:

$$H_1\text{: } \mu = \mu_0$$

possible alternative hypotheses can be:

$$H_1\text{: } \mu \neq \mu_0$$

$$H_1\text{: } \mu < \mu_0$$

$$H_1\text{: } \mu > \mu_0$$

The research question of interest will determine the direction of the alternative hypothesis. The p-value obtained for a one-tailed test will be half that of the p-value calculated for a two-tailed test.

Student's *t* test

A z-statistic may be used to gauge the disagreement between the sample mean $\bar{y}$ and the null value μ_0. In the tests conducted in the sections above the z-statistic is

$$z = \frac{\bar{y} - \mu_0}{\sigma / \sqrt{n}}$$

Assuming that the data in the population are normally distributed, the sample mean is also normally distributed, and a p-value may now be obtained from Table A1 in Appendix A. If the test is two-tailed it will be necessary to double the p-value obtained from the table.

In practice this method is rarely useful, because in most cases you do not know the value of the population standard deviation, σ. When σ is replaced by s, the sample standard deviation, the formula becomes:

$$t = \frac{\bar{y} - \mu_0}{s / \sqrt{n}} \qquad (7.4)$$

Note that the symbol z has been replaced by t, which is accordingly called a t-statistic instead of a z-statistic.

A p-value may be obtained from a t-statistic in the same way as that obtained from a z-statistic, that is, by computing the area in the two tails of a standardised distribution or in one-tail of the distribution if the alternative hypothesis indicates one of a "greater than" or "less than" direction. However, for a t-statistic, the appropriate standardised distribution is not the normal distribution. It is another distribution, called *Student's t* distribution because its inventor, W.S. Gossett, wrote under the pseudonym 'Student'.

The t distribution is graphed in Figure 7.6 and its two-tailed p-values are tabulated in Table A2 of Appendix A for various magnitudes of t.

Student's t distribution is more complex than the standardised normal distribution, because it depends on the sample size, n. Since n must be two or more to define a sample standard deviation s, the t distribution is expressed in terms of $n-1$ rather than n, where $\nu = n-1$ is called the number of *degrees of freedom*.

Note that, in the limit as ν (or n) tends to infinity, the t distribution approaches the standardised normal distribution.

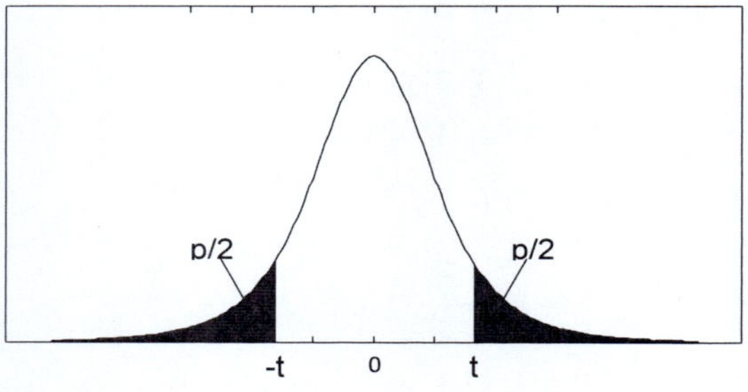

Figure 7.6: Tail Area of Student's t Distribution for a two-tail test

For brevity Table A2 in Appendix A gives values of t corresponding to selected p-values (0.0005, 0.001, 0.002, 0.005, 0.01, 0.02, 0.05, 0.1, 0.2, and 0.5) and for values of ν from 1 to 30, 35, 40, 45, 50, 60, 70, 80, 90 and 100. The bottom line of the table gives the limiting values of t when ν tends to infinity. Since the standardised normal distribution arises as the limit of the t distribution when n tends to infinity, these values are the same as those obtained from the standardised normal distribution given in Table A1 in Appendix A.

As an illustration of the *t* test, consider the body fat percentages of the subjects graphed in Figure 2.2. According to one medical authority, normal adult men are one quarter fat. Expressing this statement as a null hypothesis, μ_0 is 25 so the null and alternative hypotheses are

$$H_0: \mu = 25$$

$$H_1: \mu \neq 25$$

The body fat percentages for the four males in the study are 7.8, 9.5, 17.8 and 27.4. Here there is no indication that body fat percentages for males may be more or less than 25, so we therefore perform a two-tailed test. What is the p-value obtained from testing the null hypothesis based on these data?

Now the sample size is $n = 4$, the sample mean is 15.62, and the sample standard deviation is 8.984, so the *t*-statistic is

$$t = \frac{\bar{y} - \mu_0}{s / \sqrt{n}} = \frac{15.62 - 25}{8.984 / \sqrt{4}} = -2.088$$

The number of degrees of freedom is $\nu = n - 1 = 3$. Refer to the table of *t* values shown in Table A2. When looking at two tails of the distribution the *t* value with 3 degrees of freedom is 2.353 when p = 0.10 and is 1.638 when p = 0.20, so the p-value is between 0.1 and 0.2. (In this case the precise p-value is 0.128 (using a computer package.)

What is your conclusion? Since the p-value is greater than 0.05, it is not correct to reject the null hypothesis is this case. But the sample size is only 4, so the result is not very conclusive. You certainly cannot accept the null hypothesis. All you can say is that the mean body fat content of the males in the target population 'could be' 25%.

P-values, by themselves, are not particularly informative. It is always better to supplement a p-value by showing a confidence interval. As you saw in the preceding chapter, a 95% confidence interval for a population mean, when the population standard deviation is also unknown, is given by the formula

$$\left(\bar{y} - t_\nu \frac{s}{\sqrt{n}}, \bar{y} + t_\nu \frac{s}{\sqrt{n}}\right)$$

In this formula, t_ν is the critical value of the t-statistic with ν degrees of freedom corresponding to a two-tailed area of 0.05. The values of t_ν for various values of ν are thus given as the column of Table A2 headed by 0.05 for the two tailed alternative. For example, when n is 4 or ν is 3, t_ν is 3.182.

Substituting $\bar{y}$ = 15.62, s = 8.984, n = 4, and substituting t_ν = 3.182 into the formula, a 95% confidence interval for μ is thus

$$\left(15.62-3.182\frac{8.984}{\sqrt{4}},\ 15.62+3.182\frac{8.984}{\sqrt{4}}\right) = (1.33, 29.91)$$

This confidence interval is graphed in Figure 7.7, together with the hypothesised value of 25%. From this graph it is clear that the result is not very conclusive because the confidence interval covers a very wide range.

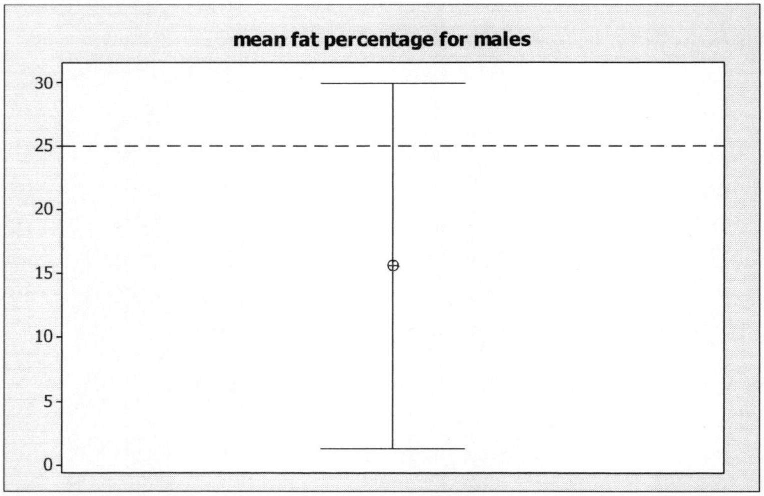

Figure 7.7: 95% confidence interval for mean fat percentage for males

***Exercise 7.2**:*

The 14 women in the sample had a mean fat percentage of 32.32, with standard deviation 4.90. Construct a graph of a 95% confidence interval for the corresponding population mean. Use these data to determine the p-values associated with each of the following null hypotheses:

a. women are 25% fat
b. women are one third fat.
c. obtain the p-values associated with parts a. and b. above but this time use an alternative hypothesis of "greater than".

Testing that a population has a specified proportion

In the preceding section you saw how a p-value is a probability associated with a null hypothesis, based on a sample of data. We illustrated this statistical concept using the mean of a sample calculated from a specified measurement, where the null hypothesis takes the form

$$H_0: \mu = \mu_0$$

The p-value is calculated as the area in the two tails (or one tail if we are conducting a one-tailed test) of the standardised normal distribution beyond the value $|z|$, based on the z-statistic

$$z = \frac{\bar{y} - \mu_0}{\sigma / \sqrt{n}}$$

where $\bar{y}$ is the sample mean, n is the sample size and we have a known standard deviation.

The same reasoning may be used to test a hypothesis involving the proportion of a population, π, that has a specified attribute. In this case the null hypothesis takes the form

$$H_0: \pi = \pi_0 \qquad (7.5)$$

A z-statistic may be used to gauge the disagreement between the sample proportion and the null value π_0. As in the case of a mean, this statistic is obtained by dividing the difference between the two quantities by the standard error of the sample proportion. Using the formula given in

Chapter 5 (Equation 5.6), the standard error of a sample proportion is $\sqrt{\pi(1-\pi)/n}$ where π is the population proportion, so the z-statistic is

$$z = \frac{p-\pi_0}{\sqrt{\pi(1-\pi)/n}} \tag{7.6}$$

The only problem with Equation 7.6 is that it involves the unknown population proportion π. You may recall that a similar problem arises when finding a formula for a confidence interval for π, as you saw in the preceding chapter.

The solution in that case is to replace π in the formula by p, its estimate from the sample. In the present case, given that we have a hypothesised value for π, namely π_0, it is better to replace π by π_0 in Equation 7.6, giving the formula

$$z = \frac{p-\pi_0}{\sqrt{\pi_0(1-\pi_0)/n}} \tag{7.7}$$

Equation 7.7 is preferred because it contains a fixed value for the unknown population proportion rather than an estimate based on the sample which is subject to sampling variation and increases the error.

As indicated previously, as a rough rule of thumb, a p-value is less than 0.05 if the corresponding 95% confidence interval for the population parameter does not contain the null value, and conversely, a p-value is greater than 0.05 if the 95% confidence interval contains the null value. This rule of thumb would be exact if the standard error used to compute the confidence interval was based on the null value rather than the estimated value. (Note that for a population proportion, the standard error used to compute a confidence interval is given by $\sqrt{p(1-p)/n}$ whereas the standard error used to scale the corresponding z-statistic is given by $\sqrt{\pi_0(1-\pi_0)/n}$.)

As an illustration of the z test for a proportion, suppose a car insurance company states in its annual report that the overall proportion of claimants in the past, up until a year ago, has been 25%. If there was no change in the propensity of insurers to make claims, it would be reasonable to hypothesise that the proportion of claimants in the most recent 12-month period has remained at 25%.

Expressing this statement as a null hypothesis π_0 is 0.25 so the null and alternative hypotheses are

$$H_0: \pi = 0.25$$

$$H_1: \pi \neq 0.25$$

Note that there is no suggestion of any direction for the alternative hypothesis and hence if is one of "not equal to" in this case. After taking a sample of 400 current insurers you find that 94 made a claim in the last year. What is the p-value obtained from testing this null hypothesis based on these data?

Now the sample size is 400 and the sample proportion is 0.235, so the appropriate z-statistic is

$$z = \frac{0.235 - 0.25}{\sqrt{0.25(1 - 0.25) / 400}} = \frac{-0.015}{0.0217} = -0.69$$

Referring to the normal table, Table A1 in Appendix A, the p-value is $0.2451 \times 2 = 0.4902$.

So in this case we would not reject the null hypothesis and conclude that the proportion of claimants in the population could be 0.25.

Exercise 7.3:

a. *A company claims that it does not take gender into account when it hires employees. To test this claim, it randomly samples nine employees from those hired during the last year, and finds that four were women and five were men. Compute a p-value. Does the size of this p-value demonstrate that the company's claim is correct? Explain.*
b. *How big a difference is needed between the numbers of women and men hired in the sample of nine to refute the company's claim?*
c. *If you suspected that the company discriminated with respect to gender in its hiring policy, would you accept the result based on a sample of nine? Explain.*

The normality assumption

There is an important assumption that underlies the methods of statistical hypothesis testing covered in this chapter, namely, that the quantity being estimated has a *normal* distribution.

When the z- statistic is used to test the null hypothesis given by Equation 7.7, it is assumed that, if repeated samples of the same size n were taken under the same circumstances, each giving a sample proportion, then the distribution of these sample proportions would be normal. This is true only approximately.

In fact the sample size n is not the most important factor determining the validity of the normal approximation when testing a null hypothesis involving a population proportion. Rather, it is the minimum of the quantities $n\pi_0$ and $n(1-\pi_0)$. As a rule of thumb, this minimum should be at least 5, otherwise the normality assumption is not reasonable.

Note that in the example given above we took π_0 to be 0.25 and $n = 400$, so the minimum is 100, and the normality assumption is justified.

The other null hypothesis we considered involves a population mean (Equation 7.4). In this case the t-statistic is used to get a p-value, and it is the sample mean that is required to have a normal distribution when repeated samples are taken. Now if the data themselves are sampled from a normal distribution, the sample mean will also have a normal distribution, so there is no problem. But if the data do not arise from a normally distributed population, the resulting p-value is only approximate.

If the distribution generating the data is a good approximation to a normal distribution, then the resulting p-value is approximately correct. But if this distribution is particularly skewed or otherwise non-normal, the p-value is not valid.

How can you decide whether the data could have come from a normal distribution, or at least from an approximately normal distribution? You could draw a histogram of the data and check that it looks relatively normal. But perhaps the best approach is to look at plots of normal scores of the data, as shown in Figures 4.13 - 4.17. If a normal scores plot follows a straight line apart from variations that could be due to sampling fluctuations, you are entitled to conclude that the normality assumption is reasonable.

Figure 7.8 shows normal scores plots for the body fat percentages of the 4 men (left panel) and for the 14 women (right panel). In each case, there is no justification for rejecting the normality assumption.

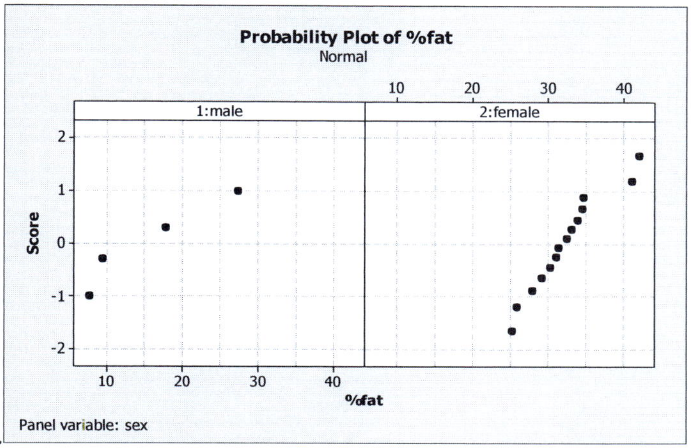

Figure 7.8: Normal scores plots for men's & women's body fat percentages

It is sometimes claimed that, if a sample is too small, it is unreasonable to assume normality because there is insufficient evidence. This claim does not follow scientific convention. In practice, normality may be assumed unless there is evidence to the contrary.

The following data, cited in Hand et al (1994, page 118), provide an illustration where the normality assumption is dubious. These are the breadth-to-length ratios of beaded rectangles used by the Shoshoni American Indians to decorate their leather goods.

0.693	0.749	0.654	0.670	0.662	0.672	0.615	0.606	0.690	0.628
0.668	0.611	0.606	0.609	0.601	0.553	0.570	0.844	0.576	0.933

Figure 7.9: Breadth-to-length ratios of 20 beaded rectangles

The authors speculated that these craftsman may have tended to make their rectangles in the shape of the "golden rectangle" which was used extensively in ancient Greek architecture, where the breadth-to-length ratio is $2{:}\sqrt{5}+1=0.618$. Figure 7.10 shows a normal scores plot of these data. This plot shows a distinct bend, so the normality assumption is not plausible. In fact there is evidence of clustering as well as skewness.

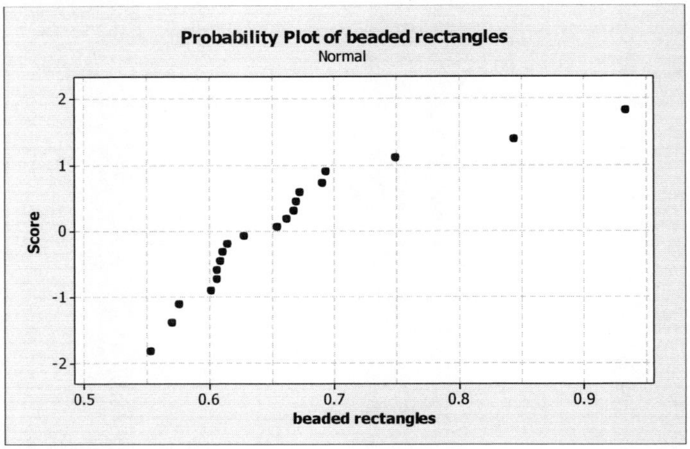

Figure 7.10: Normal scores plot for breadth-to-length ratios of Shoshoni Indian beaded rectangles, indicating a departure from normality

Real importance versus statistical significance

The term 'statistically significant' is used to describe the result of testing a null hypothesis which gives a p-value smaller than 0.05. However, a result may be statistically significant, but not of practical significance.

For example, a large study investigating risk factors for obesity among schoolchildren involved measuring their weights and heights. To ensure that the recording instruments were reliable, the measurements were taken on two separate occasions during the day. Based on 2253 children studied, it was found that the average weight was 41.175 kilograms on the first occasion and 41.188 kilograms on the second occasion, and that the standard deviation of these increases was 0.125 kg.

A t test may be used to get a p-value based on the null hypothesis that the mean increase in the target population is 0. The mean increase in the study sample is $\bar{y} = 41.188 - 41.175 = 0.013$, the standard deviation is calculated as $s = 0.125$, n is 2253, and μ_0 is 0, so Equation 7.4 gives

$$t = \frac{\bar{y} - \mu_0}{s/\sqrt{n}} = \frac{0.013 - 0}{0.125/\sqrt{2253}} = 4.94$$

The number of degrees of freedom is 2252, effectively infinity, so referring to the bottom line of Table A2, the p-value is less than 0.0005. The result is statistically significant: the null hypothesis is rejected.

Note that the increase in the mean weight is only 0.013 kg, or 13 grams. This amount is quite miniscule, and could be due to small inaccuracies in measurement, or possibly even because the children's' weights changed slightly during the day. Whatever the reason for the difference, it is clearly of no importance. This is an example of a situation in which there is a statistically significant result, but it is not of any importance. Situations like this usually arise when the sample size is large.

It is also possible for a study to fail to give a statistically significant result, even though it is of real importance. To see how this situation can arise, consider again the schoolchildren's obesity study. Suppose that a group of five girls had a mean weight of 40.56 kg recorded on the first occasion and a mean of 40.96 kg on the second occasion, the standard deviation of their increases being 0.36 kg.

In this case $\bar{y} = 40.96 - 40.56 = 0.40$, the standard deviation is $s = 0.36$, n is 5, and μ_0 is again 0, so Equation 7.6 gives

$$t = \frac{\bar{y} - \mu_0}{s / \sqrt{n}} = \frac{0.4 - 0}{0.36 / \sqrt{5}} = 2.48$$

The degrees of freedom is now 4, and Table A2 gives a p-value between 0.05 and 0.1. (More accurately, p = 0.068.) This result is not statistically significant. However, an increase of 400 grams could be of real importance, and would require some explanation. (It might be because the girls snuck out to the canteen during the study period!)

The moral is that the result of a study needs to be viewed in practical terms. Clearly it is useful to have some knowledge of the subject matter. It could be misleading to report just a p-value.

Summary

In this chapter you have seen how a p-value is used to measure the extent to which a sample of data is consistent with a specified population parameter. The specified value of the parameter is called the null value, and the specification is called the null hypothesis. If the sample is not consistent with the null hypothesis, the p-value should be small.

By convention, if the p-value is less than 0.05, you reject the null hypothesis.

But the converse is not true: if the p-value is not small, it does not follow that the null hypothesis is true. Although it is possible to reject a null hypothesis with a small amount of data, a much larger sample is needed before you can accept a null hypothesis.

We described two situations involving hypothesis testing:

a. testing that a population mean has a given value μ_0.

b. testing that a population proportion has a given value π_0

For a two-tailed test, the p-value is defined in each case as the area in the two tails of a standardised distribution. In case a. Student's *t* distribution (a more general form of the standardised normal distribution involving an additional 'degrees of freedom' parameter equal to $n-1$) is used to get the p-value. In case b. the p-value is obtained from the standardised normal distribution (and the result is only approximately valid).

To be valid, these statistical tests require assumptions about the nature of the data. The test for the population mean assumes that the population from which the data are sampled is normally distributed. The test for the population proportion assumes that both the expected number of events $n\pi_0$ and non-events $n(1-\pi_0)$ should be reasonably large, and certainly no less than 5.

When carrying out these statistical tests, the results should be shown in an informative graph, such as a confidence interval for the population parameter and a line denoting the position of the null value with respect to this confidence interval. It is also very informative to include an appropriate graph of the original data, such as a histogram or box plot.

As an illustration, Figure 7.11 gives such a concise graphical report of the analysis for the test of the null hypothesis that a population proportion is 0.1, based on a sample of 120 in which the observed proportion is 8 out of 120. Note that the expected counts based on the null hypothesis are 12 and 108 and the validity of the normal approximation may be assessed from these.

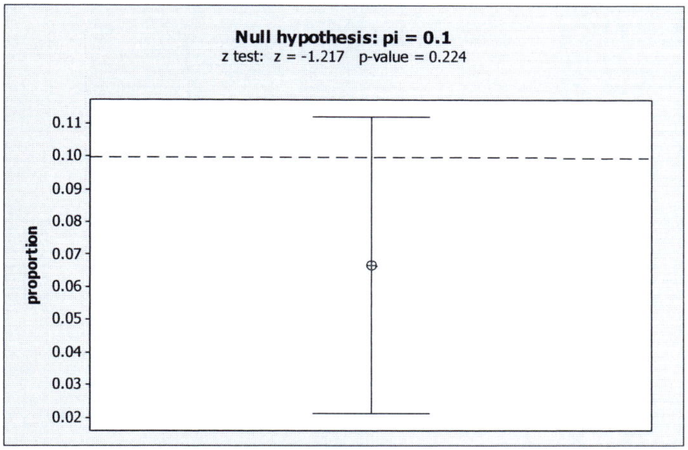

Figure 7.11: Graphical analysis of test based on a sample proportion

As a further illustration, Figure 7.12 gives a graphical report based on testing the null hypothesis that μ is the golden rectangle ratio 0.618, using the sample data listed in Figure 7.9.

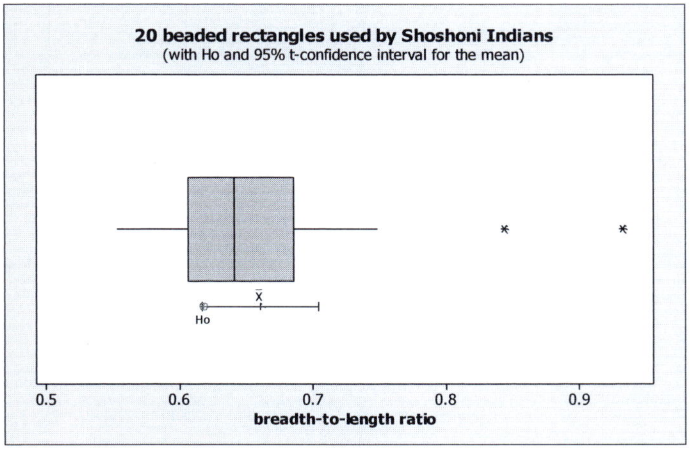

Figure 7.12: Graphical analysis of data from breadth-to-length ratios of beaded rectangles listed previously

Homework Problems

1. A data-processing company claims that its error rate for non-verified data entry is 1%. An audit reveals 113 errors among 9000 characters entered by a particular operator. Use these data to test the claim.

2. The national unemployment rate in Australia in December 1992 was 11.3% (*Sydney Morning Herald*: 7 May 1993).
   a. Of 454 persons interviewed in Western Australia's Chapman Valley in Pilbara, two were out of work. Test that this region's unemployment rate is the same as the national average.
   b. In the Queensland settlement of Karawatha, 8 of 17 persons interviewed were unemployed. Is the unemployment rate in this area statistically different from the national average?
   c. Check that any statistical assumptions you have made.

3. A United States senator, concerned about the effects of radioactive fallout on soldiers at a weapons test, asks the following question at a hearing of a government committee. "Given that 1500 soldiers who were exposed have been traced, and given that 25 of them have developed or died of a disease attributable to fallout, is this a 'statistically significant' result?" Answer the senator's question, using the additional information that the disease rate for persons in the relevant nonexposed target population is at most 1 percent.

4. The average age for patients with kidney disease in a large clinical study is 60 years. An investigator studying the psychological effects of the disease finds that 43 patients 'free of stress' have an average age of 67 years with a standard deviation of 16 years. Is this 7-year age difference statistically significant? State any assumptions you have made in answering this question.

5. Trees in a plantation have their bark thicknesses measured at 10-year intervals to determine their growth rates. The following growth rates (in mm/year) were obtained from a particular plot of *pinus radiate.*

   1 0.5 1.5 1.3 1.2 0.7 0.42 0.34 0.18 0.34 0.04 –0.02 0.26
   0.12 0.12 0.72 0.56 0.14 0.52 0 0.4 0.5 0.53 0.3 0.5 0.2 0.5

   According to theory the growth rate for *pinus radiata* trees should be 0.5 mm/year. Test this theory using the given data.

6. A bioengineering firm develops a new treatment for stimulating bone regeneration in complex tibula fractures, which involves implanting a small electrical device in the bone at the fracture site. The intervals in months from implant to complete healing are as follows:

   4 9 9 3 4 7 5 4 2 7 4 12 19 4 9 10

   A literature search reveals that such fractures normally take 9 months to heal on average, using the standard therapy. Do these data indicate that the new treatment is different to the standard therapy? Check that any assumption you have made is reasonable.

7. Solowij, Mitchie and Fox (1991) conducted a study to determine the effects of long-term cannabis use on selective attention. This study involved taking measures of brain event-relation potentials on nine users and nine non-users to ascertain their ability to process information. The subjects were instructed to listen to a particular location and pitch and to respond as rapidly as possible to the long-duration tones by pressing a button.

   It is known that the average response time for this measurement is normally distributed with mean 540 and standard deviation 80 milliseconds. The users recorded an average response time of 573.06 milliseconds. Do they differ from the general population?

8. Barnicot & Brothwell (1959) measured the maximum head breadth in millimetres of 84 ancient Etruscan male skulls, as follows:

141	148	132	138	154	142	150	146	155	158	150	140
147	148	144	150	149	145	149	158	143	141	144	144
126	140	144	142	141	140	145	135	147	146	141	136
140	146	142	137	148	154	137	139	143	140	131	143
141	149	148	135	148	152	143	144	141	143	147	146
150	132	142	142	143	153	149	146	149	138	142	149
142	137	134	144	146	147	140	142	140	137	152	145

   Assuming that modern Italian males have an average maximum head breadth of 142 millimetres, do the Etruscan skulls have a statistically different mean maximum head breadth to those of modern Italian males?

9. According to the *Sydney Morning Herald* of 1 January 1988, the incidence of babies born with spina bifida in the general population is approximately 1 in 1000. An analysis of 1,694 births from 16 IVF clinics around Australia found that between 1979 and 1986, six infants were born with this disease. Do these data indicate that the risk of the disease is different in IVF clinics?

10. Mendenhall, Ott & Larson (1974) recorded the murder rates (per 100,000) for a sample of 30 cities in southern USA for 1960 and 1970:

1960	*1970*	*1960*	*1970*	*1960*	*1970*
10.1	20.4	4.4	3.9	10.8	14.7
10.6	22.1	13.0	14.0	12.5	12.6
8.2	10.2	9.3	11.1	8.9	7.9
4.9	9.8	11.7	16.9	4.4	11.2
11.5	13.7	9.1	16.2	6.4	14.9
17.3	24.7	7.9	8.2	3.8	10.5
12.4	15.4	4.5	12.6	14.2	15.3
11.1	12.7	8.1	17.8	6.6	11.4
8.6	13.3	17.7	13.1	6.2	5.5
10.0	18.4	11.0	15.6	3.3	6.6

Did the murder rate change between 1960 and 1970?

COMPARISONS

CHAPTER 8

8 Comparisons

You have now seen two different measures of sampling variation, each of which gives an indication of the uncertainty inherent in a study due to the finiteness of the sample. In Chapter 6, a confidence interval is defined as an interval containing an unknown population parameter, such as a mean or a proportion, with a specified degree of certainty (conventionally 95%). In Chapter 7, you saw how a p-value can provide a measure of the consistency of a sample with respect to a hypothesis that a population parameter takes a specified (null) value, a small p-value (conventionally one less than 0.05) providing evidence against this null hypothesis. You also saw how a p-value is calculated for a hypothesis involving a population mean or proportion.

In the present chapter you will see how to compute p-values in some other situations. These situations involve the comparison of two population means or proportions.

Two-sample pooled *t* test

One of the examples used in the preceding chapter involves body fat percentages measured on four men and 14 women in a clinical study. These data were used to compute a p-value associated with null and alternative hypotheses of the form

$$H_0 : \mu = \mu_0$$

$$H_1 : \mu \neq \mu_0$$

where μ is the population mean and μ_0 its specified null value.

It is also of interest to test a hypothesis that two population means are equal, without specifying their common value. In fact this type of hypothesis is much more useful in practice, because it is not necessary to say precisely "what is the value of the population mean"? For example, you could test the hypothesis that men and women have the *same* mean body fat percentage (whatever it is). Similarly, you may be interested in testing the hypothesis that the value of some population parameter (such as the mean premium paid by a motorist for car insurance, or the mean daily rainfall in Singapore) is the same this year as it was last year.

The null hypothesis that two population means are equal is

$$H_0 : \mu_1 = \mu_2 \tag{8.1}$$

where μ_1 and μ_2 are the respective population means. If samples of size n_1 and n_2 are taken from the two populations, giving sample means $\bar{y}_1$ and $\bar{y}_2$, a t statistic may be used to test this null hypothesis, just as a t statistic (defined in Equation 7.4) is used to test a hypothesis for a single population mean. The two-sample t statistic takes the form

$$t = \frac{\bar{y}_1 - \bar{y}_2}{s\sqrt{\frac{1}{n_1} + \frac{1}{n_2}}} \tag{8.2}$$

In this formula, s is the *pooled* sample standard deviation, defined by subtracting the sample mean from each sample to give a set of $n_1 + n_2$ residuals, dividing the sum of the squares of these residuals by $n_1 + n_2 - 2$, and taking the square root of the result. If s_1 and s_2 denote the standard deviations of the two samples, respectively, it may be shown that the pooled sample standard deviation is given by the formula

$$s = \sqrt{\frac{(n_1 - 1)s_1^2 + (n_2 - 1)s_2^2}{n_1 + n_2 - 2}} \tag{8.3}$$

A p-value is now obtainable from the table of the two-tailed t distribution with $n_1 + n_2 - 2$ degrees of freedom. This statistical procedure is called the *two-sample pooled t test.*

As an illustration, consider again the body fat percentages of the adults from the clinical study. For the men, the sample size is $n_1 = 4$ and its mean and standard deviation are $\bar{y}_1 = 15.62$ and $s_1 = 8.98$, respectively, while for the women the corresponding quantities are $n_2 = 14$, $\bar{y}_2 = 32.32$ and $s_2 = 4.90$. From Equation 8.3, the pooled standard deviation is 5.885, so the t statistic is

$$t = \frac{15.62 - 32.32}{5.885\sqrt{\frac{1}{4} + \frac{1}{14}}} = -5.01$$

The number of degrees of freedom is 4 + 14 – 2 = 16, so the p-value obtained from Table A2 is less than 0.0005. The conclusion is that the null hypothesis is rejected: in this target population, the women have a higher percentage of body fat than men.

As a further illustration, consider the weights of two groups of anorexic girls. These data were provided by Professor Brian Everett from the University of London and are listed in Hand et al (1994, page 229). They arose from a randomised trial comparing weight gain treatments, where the girls were allocated to groups comprising a control group and a group given family therapy. The data below are the weights in kilograms before any treatment was administered, and it is of interest to compare the means in the two populations to ensure comparability of the two groups. In this case the null hypothesis is that the population means are the same.

Family therapy:	38.0	37.8	39.0	37.4	39.3	36.1	34.9	42.7	33.3
	36.5	37.0	37.3	35.2	37.9	40.8	39.0	39.6	
Controls:	36.6	40.6	41.7	33.6	35.4	40.1	39.6	34.1	36.6
	35.6	35.2	40.2	36.9	35.4	32.0	35.1	38.7	39.0
	38.2	36.2	38.8	38.3	36.1	35.2	32.8	40.4	

Figure 8.1: Weights in kilograms of two groups of anorexic girls

Box plots of these data, together with means and 95% confidence intervals, are shown in Figure 8.2. The girls in the family therapy treatment group are slightly heavier on average than those in the control group (37.75 kg compared to 37.0 kg), but the box plots substantially overlap so the difference may not be statistically significant.

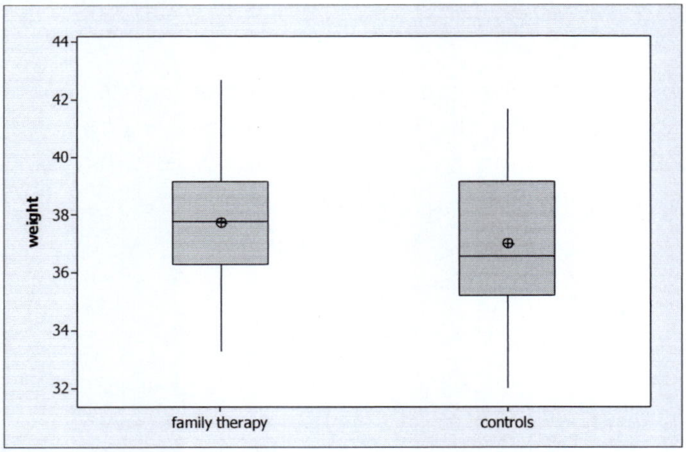

Figure 8.2: Weights of anorexic girls prior to weight gain treatment

We need a p-value for testing the null hypothesis that the population means are equal. The sample sizes are $n_1 = 26$ (controls) and $n_2 = 17$ (treated group), and the standard deviations are $s_1 = 2.59$ and $s_2 = 2.27$, so using Equation 8.3 the pooled sample standard deviation, s, is 2.47. Substituting this value into Equation 8.2, the two- sample t statistic is

$$t = \frac{37.0 - 37.75}{2.47\sqrt{\frac{1}{26} + \frac{1}{17}}} = -0.97$$

The number of degrees of freedom is 26 + 17 – 2 = 41. Referring to the table of the t distribution (Table A2), when ν is 40 and $|t|$ is 0.97, the p-value is between 0.2 and 0.5. So there is insufficient evidence to reject the null hypothesis: the means of the weights in the control and treatment populations could be the same.

Two-sample unpooled *t* test

In the previous section we pooled the two sample standard deviations to provide an estimate for the common population standard deviation. However, the underlying assumption here was that the two samples arose from populations with equal standard deviations.

As a rough rule of thumb, we can make this assumption and pool the two sample standard deviations so long as the standard deviation from one sample is not more than twice that of the other sample.

However, if this is not the case then we should not pool the standard deviations and instead our statistical procedure will be a *two-sample unpooled t test.* The two-sample unpooled t statistic takes the form

$$t = \frac{\bar{y}_1 - \bar{y}_2}{\sqrt{\frac{s_1^2}{n_1} + \frac{s_2^2}{n_2}}} \qquad (8.4)$$

which can be approximated by a *Student's t* distribution with *degrees of freedom* equal to

$$\nu = \frac{(s_1^2/n_1 + s_2^2/n_2)^2}{\frac{(s_1^2/n_1)^2}{n_1 - 1} + \frac{(s_2^2/n_2)^2}{n_2 - 1}} \qquad (8.5)$$

As an example, consider the tonnages of the 25 British merchant ships in 1907 which were reported by Floud (1973) and listed in Homework Problem 6.8. Of the 25 ships, 4 were sailing ships and 16 were steam ships, the remainder having an unknown source of power. Let's suppose we wish to determine whether steam ships have higher tonnages, on average, than sailing ships. Listed below are the tonnages of these 20 ships:

Steam:	739	970	2371	309	679	1272	3246	1904
	357	1080	1027	2507	502	1501	2750	192
Sail:	236	26	28	138				

Summary statistics for the two groups are listed below:

Power	**sample size (n)**	**mean**	**standard deviation**
Steam	16	1338	950
Sail	4	107	101

Because our research hypothesis is that steam ships have higher tonnages than sailing ships our null and alternative hypotheses are

$$H_0 : \mu_1 = \mu_2$$

$$H_0 : \mu_1 > \mu_2$$

Since the standard deviation for steam ship tonnages is more than twice that of sailing ships we cannot use the pooled two-sample *t* test to analyse these data but we can use the unpooled two-sample *t* test so long as our measurements in each group are approximately normally distributed. Using Equation 8.5, $\nu = 16.22$ which gives $\nu = 16$ when rounded down. So using Equation 8.4, the test statistic is

$$t_{16} = \frac{\bar{y}_1 - \bar{y}_2}{\sqrt{\frac{s_1^2}{n_1} + \frac{s_2^2}{n_2}}} = \frac{1338 - 107}{\sqrt{\frac{950^2}{16} + \frac{101^2}{4}}} = 5.07$$

Using Table A2 we obtain a p-value less than 0.00025 which gives us sufficient evidence to reject the null hypothesis and accept the alternative hypothesis, indicating that steam ships have higher mean tonnages than sailing ships.

Paired *t* test

When comparing two population means, the variation between observations obscures the comparison. Suppose a manufacturer of golf balls claims that a new type of golf ball goes further down the fairway when struck with a two iron than the standard golf ball used by players. How would you design a study to test this claim?

One approach would be to invite players from a golf club to participate in the study, and to randomly divide them into two groups, with members of one group given the new type of ball and the others given the standard balls. Each golfer then measures the distance travelled down the fairway by the ball after striking it with a two iron. A two-sample t test may be used to compare the mean distances.

If the improvement due to the new type of ball were only 5 metres, say, this improvement is likely to be obscured by the differences between the players, which could be as much as 50 metres. In this case a large sample of players would be needed to detect the improvement.

A better experiment would be to rank the golfers beforehand according to their hitting power, and then to form pairs containing two golfers of similar rank. One member of each pair is then asked to hit off with the new type of ball, while the other hits off with a standard ball. You then compute the difference between the distances travelled.

The advantage of the pairing is that it eliminates from consideration the variation between golfers with differing hitting power, making it easier to detect a difference between the distances travelled by the two types of ball.

You could go further, and eliminate from consideration all variation between individual golfers, by asking each golfer to hit off twice in succession, the first time using one type of ball and the second time using the other type. This type of study design is called a *crossover design*, and is often used in clinical trials comparing two drugs.

Continuously varying paired response data may be analysed using a method called the *paired t test*. In this situation there are n pairs of observations of the form (y_1 , y_2). The null hypothesis takes the form

$$H_0 : \mu_1 - \mu_2 = 0 \quad \text{or} \quad H_0 : \mu_d = 0 \qquad (8.6)$$

where μ_1 and μ_2 are the population means corresponding to y_1 and y_2 and μ_d is the population mean corresponding to y_d, the population mean of the differences. (Note that this null hypothesis is essentially the same as the null hypothesis 8.1 tested by the two-sample t test.) The formula for the paired t test statistic is

$$t = \frac{\bar{y}_1 - \bar{y}_2}{s_d / \sqrt{n}} \quad \text{or} \quad t = \frac{\bar{y}_d}{s_d / \sqrt{n}} \tag{8.7}$$

where s_d is the standard deviation of the sample of differences $y_1 - y_2$. The degrees of freedom is $\nu = n-1$. This t statistic is the special case, when $\mu_0 = 0$, of the t statistic given by Equation 7.4 for testing the null hypothesis that a population mean has a specified value μ_0 .

Patient	before	after
1	130	125
2	122	121
3	124	121
4	104	106
5	112	101
6	101	85
7	121	98
8	124	105
9	115	103
10	102	98
11	98	90
12	119	98
13	106	110
14	107	103
15	100	82

Figure 8.3: Diastolic blood pressure measurements (mm Hg) from 15 hypertensive patients

As an illustration of the paired t test procedure, consider a study involving 15 hypertensive patients reported by McGregor et al (1979) and analysed by Cox & Snell (1981). Each subject's diastolic blood pressure was measured before and two hours after treatment with captopril, a drug designed to reduce high blood pressure. The data, expressed in mm Hg, are listed in Figure 8.3.

These data are plotted in Figure 8.4, with the initial measurements plotted horizontally and the follow-up measurements plotted on the vertical axis. The vertical lines drawn from each point to the line $y = x$ thus denote the changes. If a line goes above the line $y = x$, it means that the increase is positive, whereas the lines extending below the line $y = x$ represent decreases in diastolic blood pressure.

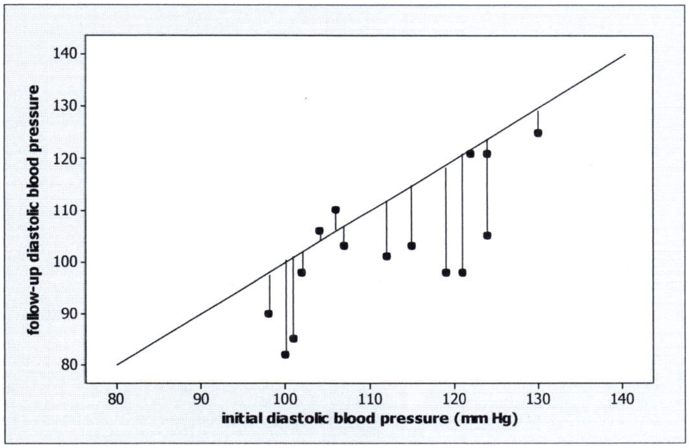

Figure 8.4: Diastolic blood pressure changes for 15 hypertensive patients

Since the researchers were expecting there to be a decrease in diastolic blood pressure we will set the alternative hypothesis as

$$H_1 : \mu_d > 0$$

where $y_d = y_1 - y_2$.

You can see from the graph that most of the changes are decreases, suggesting that the treatment is effective. The differences are 5, 1, 3, –2, 11, 16, 23, 19, 12, 4, 8, 21, –4, 4, and 18. The standard deviation of this sample of differences is $s_d = 8.6145$. The sample means are $\bar{y}_1 = 112.33$ and $\bar{y}_2 = 103.07$, respectively, so Equation 8.7 gives

$$t = \frac{112.33 - 103.07}{8.6145 / \sqrt{15}} = 4.16.$$

The number of degrees of freedom is 14, so the *t* table (Table A2) gives a p-value between 0.00025 and 0.0005 for the one-tailed alternative. The exact p-value, obtained from a computer package, is 0.00048.

Since the p-value is very small the null hypothesis (that the mean diastolic blood pressure in the target population before and after the treatment is the same) should be rejected. This statistical conclusion provides a confirmation of the evidence suggested in Figure 8.4. You can say that the mean diastolic blood pressure of a hypertensive patient is reduced by the captopril treatment.

Exercise 8.1:

a. *Compute a 95% confidence interval for the reduction in diastolic blood pressure, using Formula 6.5.*

b. *Suppose you treat the two samples of diastolic blood pressure measurements as if they were taken from two groups of 15 patients. Use the two-sample t statistic to obtain a p-value, and compare your result with that based on the paired t test.*

c. *Suppose you are given two samples of data of the same size, and asked to compare their population means. The data are arranged as in the body of Figure 8.3, so you suspect that the data are paired, but are not sure. What should you do?*

Comparison of two proportions

If the data are binary rather than continuously varying, we use proportions rather than means to summarise the results. So a method is needed for comparing two proportions. The null hypothesis of interest in this case is

$$H_0 : \pi_1 = \pi_2 \qquad (8.8)$$

where π_1 and π_2 are the respective population proportions.

This null hypothesis is the two-sample generalisation of the null hypothesis 7.5. In that case we used a *z* statistic, defined in Equation 7.7, to obtain a p-value.

In the two sample situation a z statistic is still appropriate, and is given by the formula

$$z = \frac{p_1 - p_2}{\sqrt{\bar{p}(1-\bar{p})}\sqrt{\frac{1}{n_1} + \frac{1}{n_2}}} \tag{8.9}$$

where $\bar{p}$ is the pooled sample proportion, defined as

$$\bar{p} = \frac{n_1 p_1 + n_2 p_2}{n_1 + n_2} \tag{8.10}$$

When the null hypothesis is true, the distribution of z is approximately normal. The approximation improves as n_1 and n_2 increase, provided p_1 and p_2 are not too small. A practical rule of thumb to use is that each of the four quantities $n_1\bar{p}$, $n_2\bar{p}$, $n_1(1-\bar{p})$ and $n_2(1-\bar{p})$ should be at least 5. (These are called the 'expected' numbers of events and non-events in the two samples.)

Let us apply this procedure to some data from the study of knee injuries among skiers reported by Mei Wang in her Masters thesis at Macquarie University in 1995. A question of interest is 'do women have more severe injuries than men?' Of 505 knee injuries reported at Perisher Resort in the season, 197 were sustained by men and 308 by women. Among the men, 40 injuries were classified as serious, compared to 75 serious injuries among the women.

The null hypothesis is that the proportion of serious injuries, among all knee injuries, is the same in the populations of men and women. Note that this is not the same as saying that men and women have the same risk of injury. Since we don't know the relative proportions of men and women who were skiing during the season we can't compare these risks. But it certainly appears that women are at greater risk of sustaining a knee injury while skiing, given that they had at least 50% more knee injuries and it is unlikely that there are more women than men who go skiing. Since the question of interest was one of whether women have more serious injuries than men our alternative hypothesis is:

$$H_1 : \pi_1 < \pi_2$$

From Equation 8.10, the overall proportion of serious injuries is

$$\bar{p} = \frac{40 + 75}{197 + 308} = 0.228$$

So the proportions of serious injuries among the men and the women are $p_1 = 40/197 = 0.203$, and $p_2 = 75/308 = 0.244$, respectively. Substituting these results into Equation 8.9 gives

$$z = \frac{0.203 - 0.244}{\sqrt{0.228(1 - 0.228)}\sqrt{\frac{1}{197} + \frac{1}{308}}} = -1.071$$

A p-value may now be obtained from the standardised normal distribution. From Table A1, when $|z|$ is 1.07, the p-value is 0.1423 for a one-tailed alternative. The conclusion is that there is insufficient evidence to reject the null hypothesis: among skiers who suffer knee injuries, the risk of a serious injury could be same for men and women.

Exercise 8.2:

In a study reported by Hampton & Hill (1978), 264 persons who had suffered a mild heart attack were randomly divided into two treatment groups (either hospital treatment or home management). Of the 132 sent to hospital, 14 died within six weeks, while there were 17 deaths in the same period among those looked after at home. What is your conclusion?

Matched comparison of two proportions

So far we have considered three methods for comparing the outcomes in two groups. If the outcomes are measured on a continuous scale, and the population means need to be compared, the two-sample t test (Equation 8.2) may be used. If the outcomes are binary (having one or other of two attributes), it is appropriate to compare the proportions having a given outcome in the two populations. In this case, a p-value is obtained from a z statistic (Equation 8.9) using the normal distribution.

The third method arises when the outcomes are continuous, but are matched in pairs. In this case the paired t test is appropriate and Equation 8.7 is used.

A fourth situation arises when you have matched pairs of binary outcomes. Although this design is less common as the other three, there are situations in which it is the appropriate method, as in the following example.

Sleigh et al (1982) reported a study to compare two methods (Kato-Katz and Bell) for detecting eggs from faeces samples. Each of 315 infected samples was divided into two halves, and the halves were tested using the different methods, giving the results shown below in Figure 8.5.

		Kato-Katz method +	*Kato-Katz method* –
Bell	+	184	54
method	–	14	63

Figure 8.5: Comparison of egg detection methods

Now you can see that the Kato-Katz method is successful in a proportion p_1 = 198/315, or 62.9% of the samples, while the Bell method is successful in a proportion p_2 = 238/315 (75.6%) of the samples. You could compare the corresponding population proportions using the two sample test given in Equation 8.9, but this procedure is not appropriate because the two proportions are based on paired samples, just as the two sample *t* test is inappropriate for paired continuous data.

The null hypothesis in this situation states that there is no difference between the population proportions, namely

$$H_0 : \pi_1 - \pi_2 = 0$$

and the alternative hypothesis is

$$H_1 : \pi_1 - \pi_2 \neq 0$$

(Note that this is the same as for the non-matched design, given by Equation 8.9.) This hypothesis may be tested using *McNemar's test* (McNemar, 1947). This test uses a *z*-statistic given by

$$z = \frac{m_1 - m_2}{\sqrt{m_1 + m_2}} \qquad (8.11)$$

where m_1 and m_2 are the numbers of *discordant* pairs of each type, that is, pairs for which a different outcome is obtained for the two members. Thus, for example, if the two outcomes are denoted by 1 and 0, m_1 is the number of pairs of the form (1, 0) and m_2 is the number of the form (0, 1). Note that the numbers of concordant pairs, or pairs of the form (1, 1) or (0, 0), do not enter into the formula for the z statistic.

A p-value is obtained from the standardised normal distribution in the usual way. This p-value is approximate, with its accuracy improving as the total number of discordant pairs $m_1 + m_2$ increases.

Consider the data from Figure 8.5. Denoting the outcome from the Bell method as the first member of each matched pair of results, the number of discordant pairs of the form (+, –) is $m_1 = 54$, while the number of pairs of the form (–, +) is $m_2 = 14$. Substituting these values into Equation 8.11 gives

$$z = \frac{54 - 14}{\sqrt{54 + 14}} = 4.851.$$

Referring to Table A1, the p-value is smaller than 0.000064, a highly significant result: it is reasonable to reject the null hypothesis and conclude that the Bell method is more sensitive.

If a matched case-control study design is used, the matched pairs contain one outcome of each type rather than one determinant (treatment or exposure) of each type, so the discordant pairs refer to discordant determinants rather than discordant outcomes. Despite this reversal of the role of determinant and outcome in the design, McNemar's test still applies. Consider the following example of a matched case-control design.

A study was undertaken by Donovan et al (1984) to investigate a possible link between exposure to a chemical defoliant ('agent orange') by Australian soldiers serving in the Vietnam War and the presence of birth anomalies in children they fathered after returning from the war zone.

The study design involved selecting babies born with serious genetic anomalies (the cases) from a target population, and matching them with normal babies (controls) born in the same hospital in the same month to a woman in the same age group with the same medical insurance status. The exposure status of the fathers (Vietnam veteran or not) was then measured.

The data are shown next in Figure 8.6.

		baby with genetic anomaly (case)	
		Vietnam service	*no Vietnam service*
normal	*Vietnam service*	2	121
baby (control)	*no Vietnam service*	125	8254

Figure 8.6: Data from 'agent orange' study

There are m_1 = 121 discordant case-control pairs in which the case was not exposed but the control was, and m_2 = 125 discordant case-control pairs in which the case was exposed and the control was not. Substituting these numbers in Equation 8.11 gives

$$z = \frac{121 - 125}{\sqrt{121 + 125}} = -0.255.$$

Since the p-value corresponding to $|z|$ = 0.26 is relatively large (0.7948, using Table A1), the authors thus stated that there was insufficient evidence to conclude that the risk of genetic anomalies is increased after exposure to the defoliant.

Despite the negative result, the study may not have been conclusive. What is needed in this case is a confidence interval for the increased risk associated with the exposure. If this confidence interval has a narrow range including the null value associated with no increase in risk, then it is reasonable to conclude that there is no association between the exposure and the outcome. However if the confidence interval includes values associated with a substantial increase in risk, the result remains inconclusive.

A useful measure of the increased risk of an outcome associated with a categorical determinant is the odds ratio. Odds ratios and their confidence intervals are defined in Chapter 10.

Assumptions

There are statistical assumptions underlying each of the procedures covered in this chapter. It is important to check that these assumptions are reasonable otherwise the conclusions may not be valid.

The two-sample pooled *t* test applies to the situation when you have two samples of continuously varying responses, and each observation is independent of each other observation. There are two assumptions, as follows:

a. the two populations from which the samples are drawn have the same spread;
b. the two populations are normally distributed.

These assumptions may be assessed graphically. Box plots are useful for comparing spreads because they explicitly show the midspread (the distance between the quartiles) for each sample. The normality assumption may be assessed from a normal scores plot. In this case the values plotted on the vertical axis are the pooled set of residuals rather than the original data. These residuals are obtained by subtracting each sample mean from the data in the sample and combining these mean-corrected observations. If the normal scores plot looks linear, it is reasonable to assume that assumption (b) is satisfied.

The prior weights of the anorexic girls provide a good illustration of these graphical techniques. Figure 8.2 shows box plots of the two samples, from which you can see that the sample of girls in the control group has a larger midspread than the sample who get treatment. However, the difference may not be sufficiently large to invalidate the equal-spreads assumption.

If the equal-spreads assumption is violated we would need to perform a two-sample unpooled t test using Equation 8.4.

Turning to the normality assumption, a histogram of the residuals can be used to check this assumption. However, a normal scores plot is preferable and Figure 8.7 shows a normal scores plot of the residuals. The points in this plot are extremely linear, so the normality assumption is perfectly reasonable for these data.

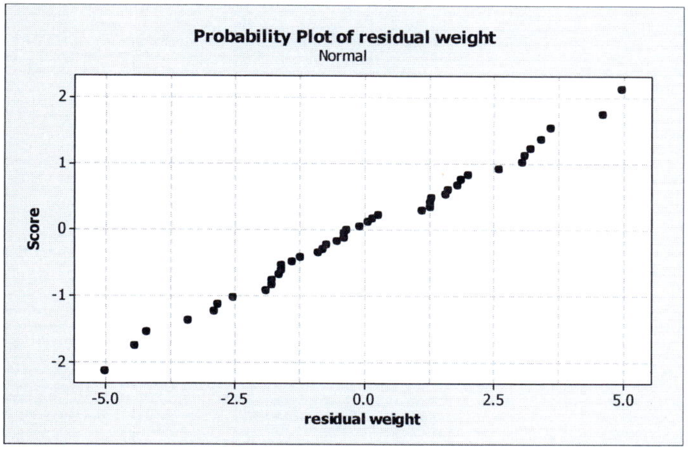

Figure 8.7: Normal scores plot of residual prior weights: 43 anorexic girls

Exercise 8.3:

Check the statistical assumptions underlying the two sample pooled t test in the case of the comparison of the mean body fat percentages for the men and women in the clinical study.

The paired t test applies to the situation when you have two samples of continuously varying responses, and the observations are paired. There is only one assumption, that is:

The differences $y_1 - y_2$ are sampled from a normal distribution.

Note that it is not required that the sample values themselves are drawn from normal populations: the assumption is only concerned with the differences. As before, a normal scores plot may be used to assess this assumption.

Figure 8.8 shows a normal scores plot of the diastolic blood pressure decreases, listed in Figure 8.3. This plot shows that the normality assumption is reasonable for these data.

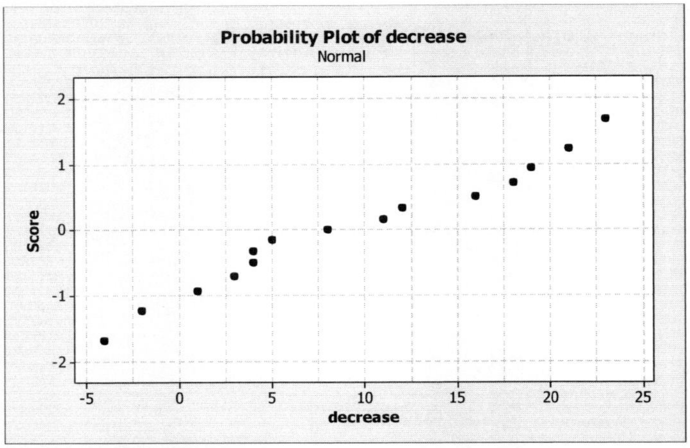

Figure 8.8: Normal scores plot of blood pressure decreases for 15 hypertensive patients

Normal scores plots also provide a method for assessing the equal spreads assumption, for the simple reason that the vertical range of the points in a normal scores plot is a measure of the spread of a sample. So, if you superimpose the normal scores plots of two samples on the same horizontal axis, you can compare their spreads directly.

Figure 8.9 gives an example of the use of a normal scores plot to compare spreads, using the prior weights of the two groups of anorexic girls. You can see from this graph that there is very little difference between the spreads of the two samples. As you saw from the box plots in Figure 8.2, the midspread of the weights of the girls in the family therapy group is less than the midspread of those in the control group, and this is confirmed by Figure 8.9. But Figure 8.9 shows, more clearly than Figure 8.2, that the equal spreads assumption is not unreasonable.

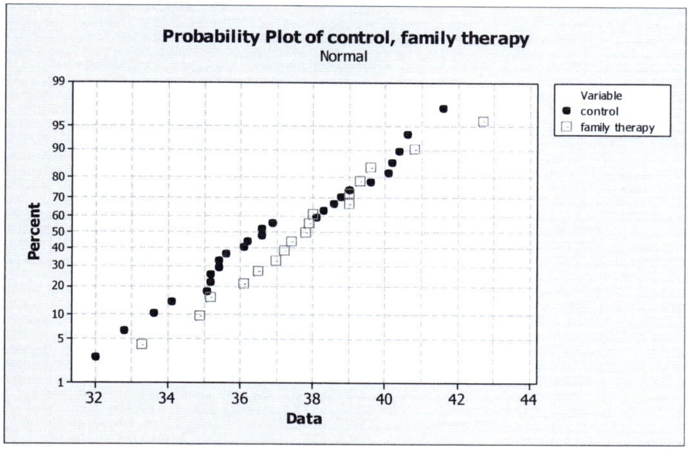

Figure 8.9: Normal scores plot of weights of two groups of anorexic girls

Summary

In this chapter the methods for computing p-values, introduced in Chapter 7 and applied to single samples of data, have been extended to four further situations involving the comparison of two population parameters.

The first situation we considered involved the comparison of two population means μ_1 and μ_2, the null hypothesis being that these population means are the same. You saw how a two-sample *t* test may be used to obtain a p-value associated with this null hypothesis based on a sample from each population. Depending on whether we believe the samples arise from populations with equal spreads will determine whether we perform a two-sample pooled or unpooled *t* test.

The second situation again involves the comparison of two population means, again based on two samples, but with the proviso that the samples are paired (and thus necessarily of the same size). In this case the appropriate method is the paired *t* test, which is really just a special case of the *t* test based on a single sample described in Chapter 7. The special case arises by using the sample of differences $y_1 - y_2$ to test the null hypothesis that the corresponding difference in population means $\mu_1 - \mu_2$ is equal to 0 or some other hypothesised value.

The other two situations involve the comparison of population proportions π_1 and π_2 from separate and paired samples, respectively. In each case the null hypothesis is that $\pi_1 = \pi_2$, and a p-value is obtained, to an approximation, as the area in the two tails of the standardised normal distribution.

As in the tests based on single samples, assumptions need to be satisfied. For the two-sample pooled *t* test, the populations generating the samples should be normally distributed and, in addition, the normal distributions should have the same standard deviation. For the two-sample unpooled *t* test we only need to satisfy the normality assumption. And for the paired *t* test, the only assumption required is that the differences $y_1 - y_2$ should constitute a sample from a normally distributed population.

Box plots may be graphed to check that the population standard deviations could be the same in the two-sample *t* test, and normal scores plots or histograms are used to assess the normality assumption.

The assumption required for the two-sample proportions test is that the expected numbers of events and non-events in the two samples should all be at least 5. The paired-proportions test requires that the total number of discordant pairs should be reasonably large.

When comparing sample means, the result of the statistical analysis may be presented in a concise graph (similar to Figures 7.11 and 7.12). For the comparison of two means, this graph may include box plots of the data in the two samples, the test statistic and the p-value for the *t* test. Figure 8.10 shows such a report for the comparison of the men's and women's body fat percentages.

Similarly, a summary is useful when comparing two population proportions. Figure 8.11 shows an appropriate graphical summary of the knee injury incidences for the men and women skiers referred to earlier in this chapter. Note that the *z* statistic and the p-value are slightly different from those given earlier in the chapter, due to rounding in the calculations. The results shown in Figure 8.11 are more precise.

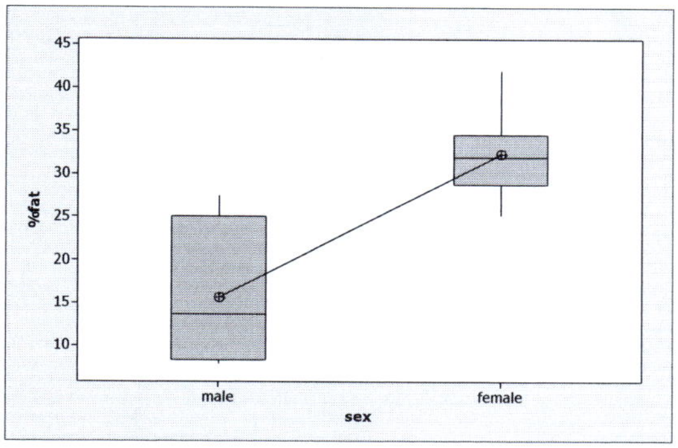

Two-Sample T-Test and CI: %fat, sex

```
Two-sample T for %fat
sex         N   Mean  StDev  SE Mean
male        4  15.63   8.98      4.5
female     14  32.32   4.90      1.3
Difference = mu (1) - mu (2)
Estimate for difference:  -16.70
95% CI for difference:  (-23.77, -9.62)
T-Test of difference = 0 (vs not =): T-Value = -5.00  P-Value = 0.000  DF = 16
Both use Pooled StDev = 5.8856
```

Figure 8.10: Graphical summary of comparison of body fat percentages

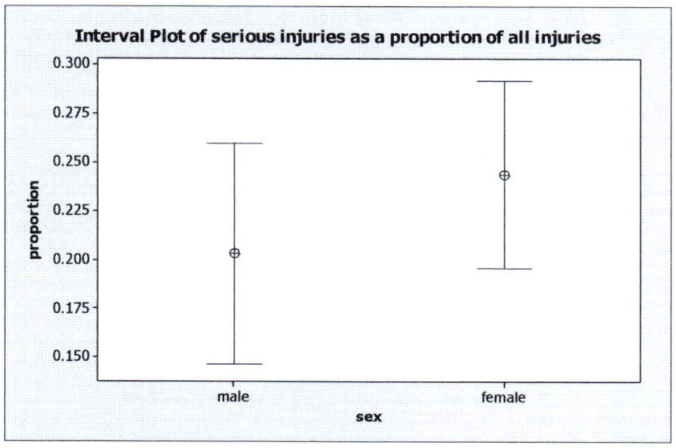

Test and CI for Two Proportions: serious, sex

```
Event = male
sex        X    N  Sample p
male      40  197  0.203046
female    75  308  0.243506
Difference = p (1) - p (2)
Estimate for difference:  -0.0404608
95% upper bound for difference:  0.0215111
Test for difference = 0 (vs < 0):  Z = -1.07  P-Value = 0.141
```

Figure 8.11: Graphical summary of comparison of skier knee injuries

Homework problems

1. The local court appearances for a number of areas in Sydney were reported in the *Sydney Morning Herald* on 20 May 1993 as rates per 100,000. Randomly selected samples from northern and southern suburbs are listed as follows:

Northern suburbs		*Southern suburbs*	
Manly	1997	Rockdale	1578
Willoughby	1091	Strathfield	1188
North Sydney	1246	Botany	1839
Warringah	1633	Hurstville	1434
Wyong	2318	Sutherland	1492
Blacktown	1925	Fairfield	1821
Lane Cove	815		

   a. Graph these data and comment on the display.

   b. Is there evidence of a difference in the court appearance rates between the northern and southern suburbs?

2. Gastwirth (1988, page 526) presented the results of a random roadside survey of young male and female drivers in Oklahoma. In this survey, of 619 drivers under 20, 77 out of 404 men and 16 out of 122 women responded that they had consumed alcohol during the previous 2 hours.

   a. Calculate the proportions of men and women drinkers.

   b. Compare these proportions using an appropriate hypothesis test.

   c. Graph 95% confidence intervals for the two population proportions, and comment on this graph in relation to the decision in part b.

3. A study conducted by Warton & Goodnow (1991) sought to determine children's attitudes to responsibility in terms of jobs around the house. Children aged 8, 11 and 14 years of age were asked to comment on the fairness of various work arrangements. The measurement recorded is the percentage of children who undertook each of 13 tasks, rather than leaving it to another family member.

   a. Is there evidence of a difference between 14 year-old boys and girls in terms of their willingness to do these tasks?

   b. Do the percentages of 8 year-old and 11 year-old girls willing to do these tasks differ?

   c. Make an appropriate graph of the data used to answer part (b). Do you think that this graph supports the conclusion reached?

	girls			*boys*		
task	*8*	*11*	*14*	*8*	*11*	*14*
Tidy own things	100	100	100	100	100	100
Make own breakfast	87	100	95	80	73	82
Make own bed	93	93	95	60	67	95
Pick up dirty clothes	80	87	91	87	67	87
Clear away dishes	87	93	95	53	73	73
Take out garbage	40	47	32	53	47	64
Do dishes/dishwasher	40	20	41	40	47	41
Hang out washing	40	33	41	60	53	50
Set table	53	33	37	80	40	27
Bring in paper/mail	47	47	32	47	53	32
Wash car	20	40	32	53	47	45
Vacuum a room	27	20	27	33	40	45
Tidy others' things	33	13	0	33	27	23

4. Grizzle, Starmer & Koch (1969) cited a clinical trial in which each of 46 subjects was given two drugs on separate occasions, where each response was recorded simply as either 'favourable' or 'unfavourable'. The results were as follows:

Favourable response for each drug:	8 subjects
Unfavourable response for each drug:	10 subjects
Favourable/Unfavourable:	20 subjects
Unfavourable/Favourable:	8 subjects

Is there any evidence of a difference between the two drugs? Answer this question after determining a p-value for an appropriate statistical test.

5. Tan & Choo (1990) undertook a study to examine the impact of deep-elaborative versus shallow-reiterative information processing on students' academic performance. The students were divided into two groups before they took their final examinations in two units (*Management Accounting* and *Accounting Information Systems*) at Macquarie University, with the following results:

Group	*Management Accounting*	*Information Systems*
Deep elaborative	n = 24, mean = 73.08 std.dev. = 6.35	n = 20, mean = 74.71 std.dev. = 6.86
Shallow reiterative	n = 24, mean = 58.62 std.dev. = 5.15	n = 21, mean = 56.20 std.dev. = 11.05

a. On the basis of these results, is there evidence of a difference between the results for the two groups?

b. For those in the 'shallow reiterative' group, is the mean score obtained by those taking the Management Accounting examination different from that obtained by those taking the Information Systems examination?

c. Is there any evidence that any of the statistical assumptions are violated?

6. Selvin (2004, Table 2.1) quoted some data from a study carried out in California among 3,154 middle-aged men in 1960-61 to investigate the relation between behaviour pattern and the risk of coronary heart disease. For the 40 heaviest men in the study, 20 were found to exhibit type A behaviour (characterised by urgency, aggression and ambition) while the other 20 exhibited type B behaviour (relaxed, non-competitive and less hurried). Their cholesterol levels, in mg/100 ml, were recorded as follows:

Type A:	233	291	312	250	246	197	268	224	239	239
	254	276	234	181	248	252	202	218	212	325
Type B:	344	185	263	246	224	212	188	250	148	169
	226	175	242	252	153	183	137	202	194	213

Among heavy middle-aged men in this target population, is cholesterol level related to behaviour type? Answer this question by first making a graph of the data and then computing a p-value for an appropriate statistical test, and then state your conclusion.

7. Hand et al (1994, page 318) have listed data from a study comparing prices of a standard (one-fifth of a US gallon) bottle of Seagram 7 Crown Whisky in different states of the USA in 1961. These states were classified according to whether the liquor stores in the state were privately owned or state monopolised. The prices in US dollars:

16 monopoly states:

4.55 4.11 4.15 4.20 4.55 3.80 4.00 4.19 4.75 4.74 4.50 4.10 4.00 5.05 5.20

26 private-ownership states:

4.82 5.29 4.89 4.95 4.55 4.90 5.25 5.30 4.29 4.85 4.54 4.75 4.85 4.85 4.50 4.75 4.79 4.85 4.79 4.95 4.66 4.75 5.20 5.10 4.80 4.29

Compare these two samples using appropriate statistical methods.

8. In the year before the introduction of random breath testing for motorists in a city the number of motor vehicle fatalities in which the driver was found to have been drinking to excess was 163. In the following year the number of such fatalities dropped to 130. Carry out a statistical test to determine whether the rate of such drink-driving fatalities changed. Assume that the population at risk has not changed. (Hint: Since the size of the population at risk is unknown, assume a large value for it. To what extent does your answer depend on the size of this population?)

9. The following table gives estimates of unemployment rates based on sample survey data for September 1992 and December 1992 in various regions in and around Sydney, as reported by the federal Department of Employment, Education and Training and published in the *Sydney Morning Herald* on 7 May 1993.

Region	*Sept 92*	*Dec 92*
Inner Sydney	11	9
Inner West	9	10
Eastern Suburbs	6	6
St George/Sutherland	5	6
Canterbury/Bankstown	13	12
Fairfield/Liverpool	17	21
Outer South-West	11	10
Central West	12	16
Outer West	8	7
Blacktown/Baulkham Hills	11	12
Lower North	5	6
Hornsby/Ku-Ring-Gai	6	8
Manly/Warringah	4	5
Gosford/Wyong	13	11
Hunter	12	13

a. Graph these data appropriately.

b. Was there a change in the unemployment rate between the September 1992 and December 1992 quarters? Answer this question using a statistical test, and give a 95% confidence interval for the change.

10. Till (1974, page 104) reported the following three samples of salinity values (in parts per thousand) taken from different water masses in the Bimini Lagoon, Bahamas.

Mass I	Mass II	Mass III
37.54	40.17	39.04
37.01	40.80	39.21
36.71	39.76	39.05
37.03	39.70	38.24
37.32	40.79	38.53
37.01	40.44	38.71
37.03	39.79	38.89
37.70	39.38	38.66
36.36		38.51
37.75		40.08
37.45		
38.85		

Compare the salinity values in these three water masses, using methods covered in Chapter 8.

11. Suppose you have two alternative routes (A or B, say) for getting from your home to your place of work, and you are not sure which route is quicker. To find the answer, you persuade a colleague who lives near you to join you in an experiment. On the first day you take Route A and your colleague takes Route B, and the next day you take Route B and your colleague takes Route A, continuing in this way for an even number of days. The following times in minutes are recorded after four days:

Day	*Route A*	*Route B*
1	37.6	39.0
2	40.5	41.5
3	39.0	39.2
4	40.7	42.6

a. Analyse the data and state your conclusion.

b. Can you see any problems with this study design? Make some suggestions for improving the efficiency of the experiment.

RELATIONS

CHAPTER 9

9 Relations

The associations between determinants and outcomes are of primary interest in many research studies. Examples include:

a. the association between gender and injury level among skiers with knee injuries;
b. the association between the administration of a drug and the diastolic blood pressure of hypertensive patients;
c. the association between the maximum highway speed limit and the accident rate in different countries;
d. the association between the atmospheric air pollution level and the incidence of bronchitis among children in a region.

In situation a. the determinant is binary (male or female gender) and the outcome is ordinal (whether the injury is mild, moderate, or severe). In situation b. the determinant is also binary (whether or not the drug is administered), but in this case the outcome varies continuously. In situations c. and d. both the determinant and the outcome are continuous variables.

In this chapter and the next we focus on graphing, measuring, and assessing associations between variables. The present chapter deals with situations in which the outcome variable is continuous, while Chapter 10 covers examples in which both variables are categorical. (Situations involving categorical outcomes and continuous determinants require, strictly speaking, more advanced methods.)

The method for investigating an association involving a continuous outcome variable is called *linear regression analysis*, and in this situation the term *relation* is used to describe the association. This method extends to a much wider range of problems, and is the basis for most of the statistical methods used in practice today.

In this chapter we start with the scatter plot, which gives a very effective visual impression of the relation between the determinant and outcome variables. Next you will encounter methods for summarising the relation based on a simple linear model, and for assessing the goodness-of-fit of this model. Finally, you will learn how to compute both confidence intervals for population parameters arising in this model, and p-values for testing the appropriate null hypothesis.

Scatter plots

Suppose you have two variables, each defined on an interval scale, and you wish to see how they are related. If both are measured on a single subject in an experimental situation one variable could be regarded as a stimulus or determinant. This determinant variable may affect the other variable, which is thus called the dependent variable or response variable. Our aim is to measure the strength of the relationship between the two variables and to find a relationship between them that will enable prediction of values of the response variable, given a value for the determinant.

It is useful to have a model or formula relating the two variables.

Examples of this situation occur in many fields of study. In medicine, you may be interested to know how a person's weight affects their blood pressure. In sport, you may wish to determine how the amount of training affects an athlete's performance. In finance, you may wish to study the effect of money supply on interest rates. And in education you may wish to see how a student's IQ affects their examination performance.

To measure the degree of association between two variables, the measurements need to be paired. So the data structure is similar to that used for the paired t test considered in Chapter 8, where you saw that the difference between the two observations within each pair contains the information of interest. (For example the decrease in diastolic blood pressure is of primary interest in the example considered in Chapter 8.)

When examining relations the focus is not on the differences, because the two variables are usually measured using different units or scales. All we have are two measurements taken from the same subject or unit.

Consider a study to examine a possible relation between the price of biscuits in a supermarket and the number of packets sold on a given day. The sample comprises ten types of biscuit (labelled A-J), and the data obtained from the supermarket in question on a given day are as listed in Figure 9.1. Bicky Streepson (1996) supplied the data. Is there a tendency for the cheaper biscuits to be sold in greater quantities, or are sales independent of price?

The first step is to examine the data visually. If you wish to study the association between the two variables you need to show both variables on the same plot. The usual convention is to make a plot of (x, y) points with the determinant (x) on the horizontal axis and the outcome (y) on the vertical axis. This plot is called a *scatter plot.*

Figure 9.2 shows a scatter plot of the data listed in Table 9.1. To assist with identification the points in the graph are labelled.

Biscuit type	*price (cents)*	*packets sold*
A	120	40
B	139	31
C	195	15
D	78	42
E	115	45
F	156	28
G	189	21
H	95	38
I	210	19
J	174	32

Figure 9.1: Sales data obtained from a supermarket on a given day

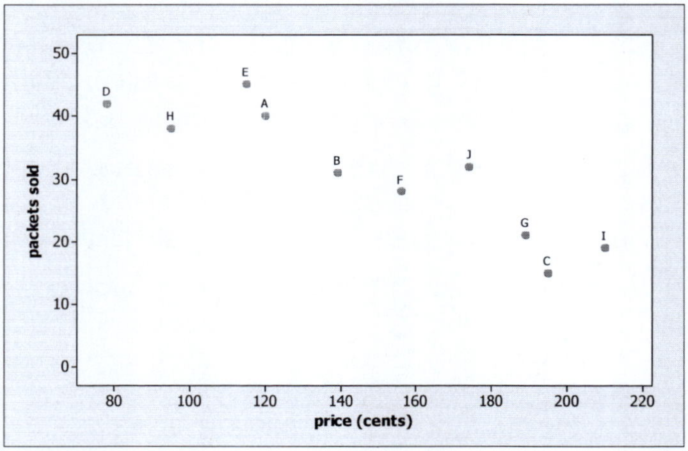

Figure 9.2: Scatter plot of biscuit packet sales versus price

When using a scatter plot to graph data like these, you should choose scales that show the data in an informative way. Note that it is not necessary to start each axis at zero, unless this can be done without undue waste of space in the graph. So Figure 9.2 has the origin for the sales volume because this is easily accommodated, but not for the packet price because doing this would create too much blank space.

It is clear from the scatter plot that the number of biscuits sold tends to decrease as their price increases. In cases like this we say that there appears to be a negative relation between the two variables.

The next graph (Figure 9.4) is a scatter plot of the fuel consumption of cars, in litres per 100 km, travelling at different average speeds, in kilometres per hour, on country roads. The data, listed in Figure 9.3, were reported by 15 graduate students as part of an assignment in a course in risk management taught by Dr Nod leMinc at Macquarie University in 1996.

Student	*speed (km/hr)*	*consumption (litres/100 km)*
Dylia	51	20
Ennjy	55	22
Cerub	60	19
Thoroyd	61	24
Heung-See	64	15
Ethik	69	23
Ixe	74	20
Ghon	80	17
KC	85	25
Hanora	90	19
Mothas	94	28
Ekhui	98	25
Mahilah	105	21
Snorly	115	27
jaK	120	30

Figure 9.3: Average speed and fuel consumption for 15 graduate students

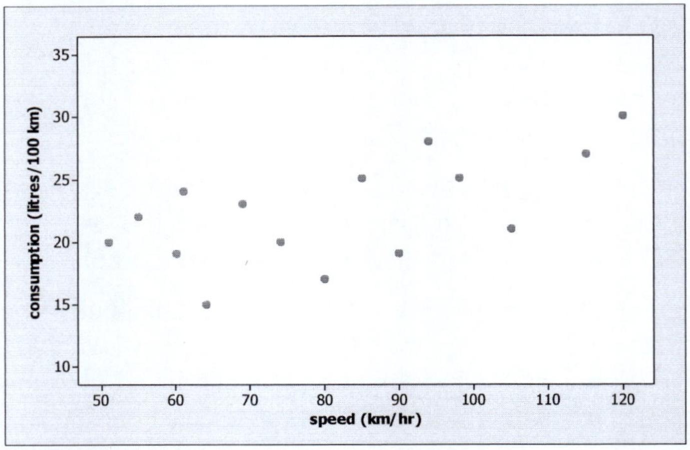

Figure 9.4: Scatter plot of fuel consumption data

As you would expect, the trend here is positive. The faster you travel the more fuel you use for the same trip. There is more variation here compared to our last plot. This is largely due to the fact that the sample contains different cars and drivers and this adds to the variability in the data.

Bivariate data also arise by repeatedly taking measurements from the same subject. In the next plot (Figure 9.6) we show the accumulated weight loss in kilograms for a trekker over a number of weeks. The data were obtained from a student (Leica Waria) who was enrolled in Stat170 in first semester 1993, obtaining a B. Elated with this result (the only grade above a C she obtained), she went trekking in Nepal from July to September 1993, and recorded her weight loss every two weeks. The data are listed in Figure 9.5.

time (weeks)	*weight loss (kg)*
2	1.8
4	2.8
6	4.2
8	4.9
10	5.3
12	5.5

Figure 9.5: Accumulated weight loss of student during Himalayan trek

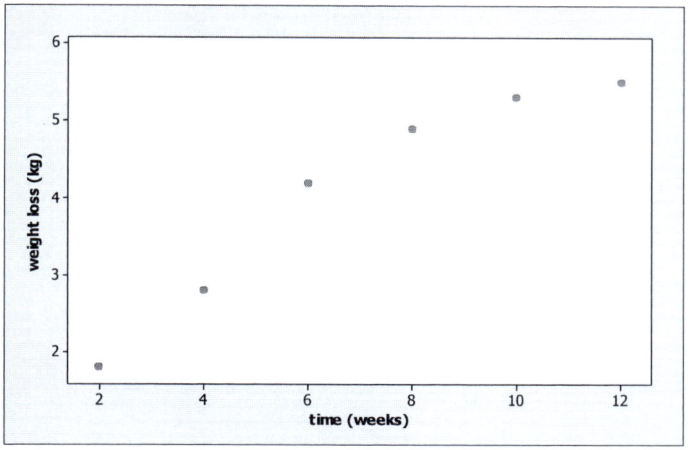

Figure 9.6: Weight loss versus time elapsed for Himalayan trekker

It is clear that the relationship is positive indicating that the longer Leica trekked the greater was her total weight loss. However it is noticeable that she did not continue to lose weight at the same rate as time progressed.

Linear relations

Scatter plots give a picture of the relationship between two variables, just as histograms and box plots do for single variables. For a sample of single measurements, you may summarise the data using a single value such as a mean, median, or proportion. The question thus arises: 'how do you summarise a set of bivariate data?'

Note that the mean or median of each component variable is not enough, because these typical values provide no information about the relation between the two variables.

To summarise the information on the scatter plot, a typical value of the outcome variable, y, is needed for each value of the determinant variable, x. Since x is assumed to take values ranging continuously over an interval on the horizontal axis, this summary gives a set of y values tracing out a curve, similar to the curve shown in Figure 9.7, using the data from the biscuit sales.

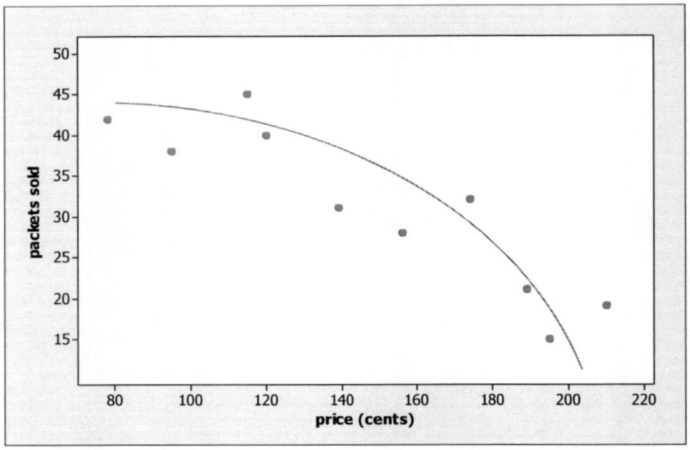

Figure 9.7: Plot of biscuit packet sales with curve summarising relation

The curve drawn in Figure 9.7 appears to summarise the relation reasonably well, but it is not the simplest curve that could be used. The simplest curve is just a straight line, like the one shown in Figure 9.8.

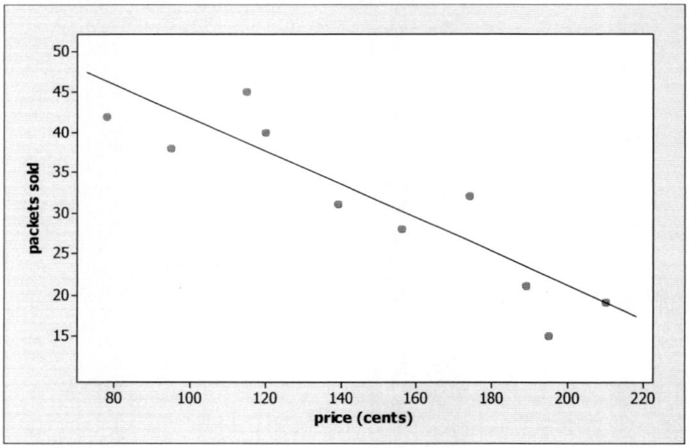

Figure 9.8: Plot of biscuit packet sales with fitted straight line

An advantage of using a straight line to summarise the relation between a determinant and a continuously varying response variable, besides being easy to draw, is that it has a very simple interpretation. It just says that there is a constant rate of increase or decrease in the response, y, corresponding to a unit increase in the determinant, x. This relation may be written as a linear equation of the form

$$y = \alpha + \beta x\,, \tag{9.1}$$

where α and β are parameters. The parameter α is called the *intercept* or *constant* and β is known as the *slope* or *gradient* of the fitted line.

A further advantage of a straight-line summary is that it only takes two points on the graph to define it uniquely. This means that you can fit a straight line to a set of points by choosing two points in the graph window that you wish the line to pass through, and then simply draw the line passing through these points.

For example, in the case of the fuel consumption data, suppose you choose the points (x = 50, y = 15) and (x = 120, y = 32) to define the fitted line. Figure 9.9 shows these points, denoted by circles, with the fitted line passing through them.

Note that you can work out the equation of the straight line passing through two points (x_1, y_1) and (x_2, y_2) using the formula

$$y = a + b\,x, \qquad (9.2)$$

where $b = (y_2 - y_1)/(x_2 - x_1)$ and $a = y_1 - b\,x_1$.

Thus if (x_1, y_1) is (50, 15) and (x_2, y_2) is (120, 32), Equation 9.2 gives us $b = (32-15)/(120-50) = 17/70 = 0.2429$ and $a = 15-(17/70)\times 50 = 2.857$, so the fitted line is

$$y = 2.857 + 0.2429\,x.$$

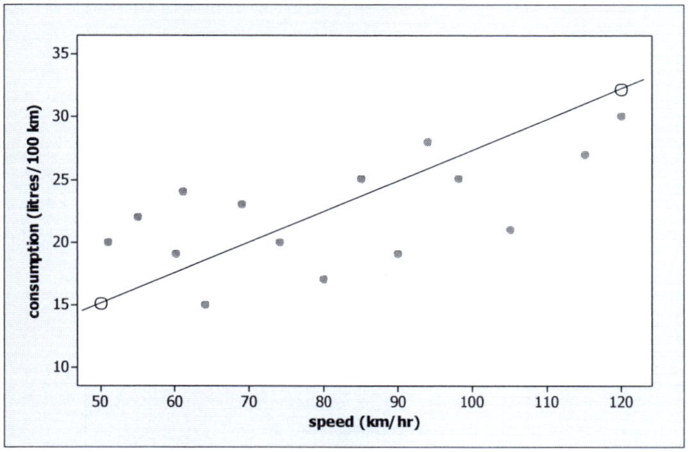

Figure 9.9: Fuel consumption data with fitted line passing through (circled) points (50, 15) and (120, 32)

Alternatively, if you are given the equation of the fitted line, you can draw it on the graph by choosing two values of x and calculating corresponding fitted values of y from Equation 9.1.

To see how this works, suppose that the equation of the straight line fitted to the biscuit sales and price data is

$$y = 61 - 0.2\,x$$

Taking the points $x_1 = 75$ and $x_2 = 225$ as a basis, the corresponding fitted values are $y_1 = 61 - 0.2\times75 = 46$ and $y_2 = 61 - 0.2\times225 = 16$. So you should draw the line to pass through the points (75, 46) and (225, 16). Note that it is better to take x_1 and x_2 to be close to the ends of the horizontal axis, to obtain greater accuracy when fitting the line by hand.

To summarise, a curve describes the relation between a determinant and an outcome variable. The simplest curve is a straight line. Our aim is to look at the sample data and determine the straight line that fits them best. Any departure from a straight line observed for points in the sample is assumed to be due to sampling variation or experimental error.

Another advantage of fitting a straight line to data is that it gives a simple method for predicting values of y using values of x which are not necessarily in the sample. For example it would be useful, based on the graph of the fuel consumption data with the fitted line, to be able to say what is the predicted fuel consumption for a driver travelling at a specified average speed. To do this, you simply substitute the given value of x into the formula, and thus calculate the predicted value of y.

As an illustration, suppose you wish to know the predicted value of the fuel consumption corresponding to an average speed of 100 km/hour, based on the straight line fitted to the data graphed in Figure 9.6. The formula for the fitted line is $y = 2.857 + 0.2429\,x$, so substituting $x = 100$, you get the predicted value $y = 2.86 + 0.2429\times100 = 27.15$.

Figure 9.10 shows the situation graphically. Starting with the value chosen, $x = 100$ for the determinant, draw a vertical line to meet the fitted line, and then draw another horizontal line from this meeting point across to the vertical axis where it corresponds to the predicted value of y.

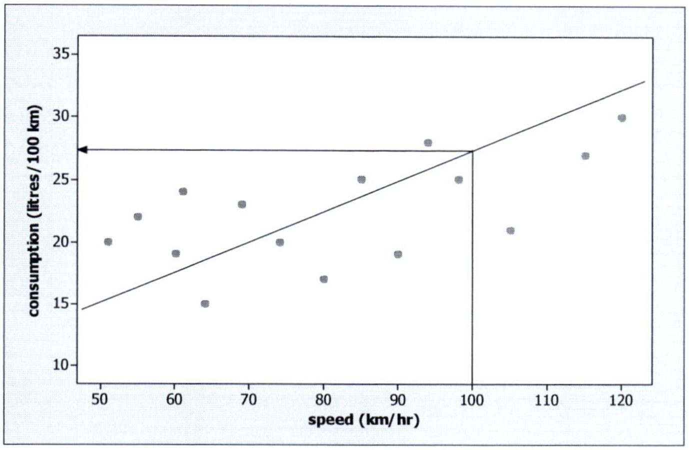

Figure 9.10: The predicted fuel consumption (27.15 litres/100km) corresponding to an average speed of 100 km/hour

Exercise 9.1:

a. *For the biscuit sales data, use the fitted line* $y = 61 - 0.2x$ *to calculate the predicted number of packets with price \$1.50 sold in a day.*

b. *Do you think it is a good idea to use a fitted line to the weight loss data to predict the trekker's weight loss after 7 weeks? Explain.*

The line fitting based on fitting by eye is subjective, and likely to give different results from different persons, and even different results by the same person on different occasions. What is needed is an objective method that gives the same answer every time.

The conventional method for fitting a straight line to a scatter plot of pairs (x, y) uses the *least squares line,* which minimises the sum of the squares of the vertical distances from the points to the line. This method is described in the next section.

The least squares line

When defining the standard deviation as a measure of spread in Chapter 3, you saw values called *residuals*: these were defined by subtracting the sample mean $\bar{y}$ from each observation, giving a new set of sample values with mean 0.

For bivariate data comprising pairs of the form (x, y) where x is a determinant and y is a response variable defined on an interval, the residuals are defined by subtracting the fitted values from the responses, that is, for the observation (x_i, y_i) the residual is

$$z_i = y_i - (a + b x_i) \tag{9.3}$$

These residuals thus correspond to the vertical distances from the points in the scatter plot to the fitted line, where the distances are positive for points above the line and negative for points below the line.

For the biscuit sales data, the residuals from the fitted line $y = 61 - 0.2x$ are shown, represented as vertical lines, in Figure 9.11. Negative residuals correspond to lines dropping beneath the fitted line, and are labelled with minus (–) signs, while the positive residuals (having points above the fitted line) are denoted by plus (+) signs. Note that the line passes through the point corresponding to x = 210, so the residual corresponding to this point is zero.

The least-squares line is the fitted line that minimises the sum of the squares of the residuals. Now it may be shown using theory that this least squares line has parameters

$$b = \frac{\sum(x_i - \bar{x})(y_i - \bar{y})}{\sum(x_i - \bar{x})^2} \tag{9.4}$$

and

$$a = \bar{y} - b\bar{x} \tag{9.5}$$

These calculations are time-consuming and error-prone if performed by hand, and it is best to obtain the estimated coefficients a and b by storing the data in a computer file and using a computer package to do the work.

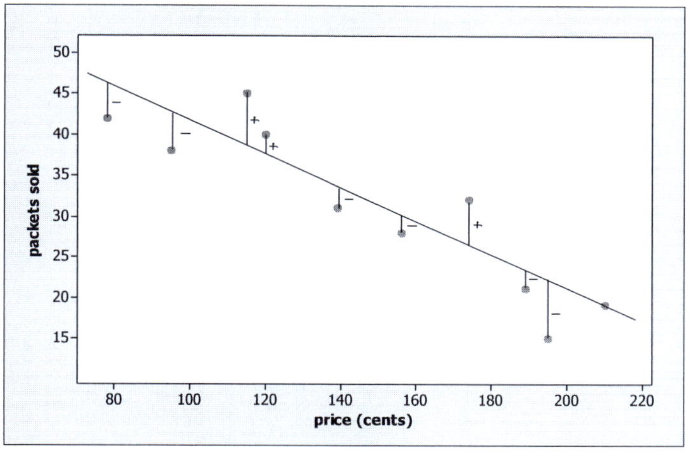

Figure 9.11: Plot of biscuit packet sales with a fitted straight line and corresponding residuals

For the biscuit data, the estimates of the intercept and the slope based on fitting the least-squares line are as follows.

$$a = 61.59, \quad b = -0.2073.$$

These estimates are thus called the *least-squares* estimates. For reasons that are steeped in obscure tradition and not very logical, the least squares estimate of the slope parameter β is called the *regression coefficient*.

Note that the least squares estimates for these data are not too different from those given before (a = 61, b = –0.2). For the fuel consumption data, the least squares line has a = 12.88 and b = 0.1162, which is rather different from the line shown before in Figure 9.9 (for which a = 2.86 and b = 0.243). Figure 9.12 shows the data with the least squares line.

As a general rule, the least squares line has a slope closer to 0 than the line you might fit to the data by eye. The least-squares line minimises the (squared) distances of the points from the line *in the vertical direction*, whereas the line you might fit by eye is closer to one that minimises the sum of (squared) distances at right angles to the line (shown as the dotted line in Figure 9.12).

The justification for using the line that minimises the error in the vertical direction is that the purpose of fitting the line is to predict the response variable, and this is measured on the vertical axis. However if there is no clear distinction between the determinant and the response variable, it might make more sense to summarise the relation using a different line (such as the dotted line in Figure 9.12).

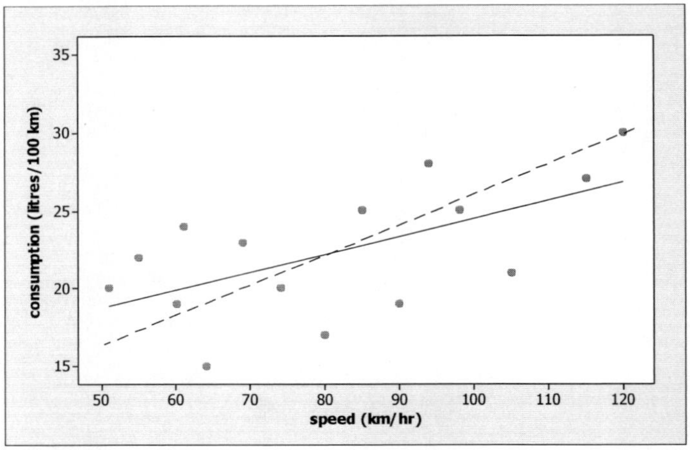

Figure 9.12: Fuel consumption data with least squares line, and line minimising sum of squared distances at right angles to the line (dotted)

The least-squares line has the useful property that it always passes through the point ($\bar{x}$, $\bar{y}$). This point is called the *centroid* of a scatter plot.

As you have seen, a fitted straight line is useful for *prediction*. Given a value of the determinant *x*, you can use it to predict the value of *y*, ***provided you do not use the fitted line to predict values outside the range of the data***.

Correlation and goodness-of-fit of line

The least squares line provides a concise summary of the relation between a determinant *x* and an outcome *y*. However, this line in itself does not tell you how well it summarises the relation.

Figure 9.13 shows two scatter plots with the least squares line fitted in each case. The fitted line is the same in each case, namely, $y = 4.73 + x$. But the plots are quite different. Even though the fitted lines are the same, the plot on the left shows much less variation around the fitted line than the right-hand plot.

The situation is similar to the two samples of students' ages given in Chapter 3, where the sample means are the same but the spreads differ.

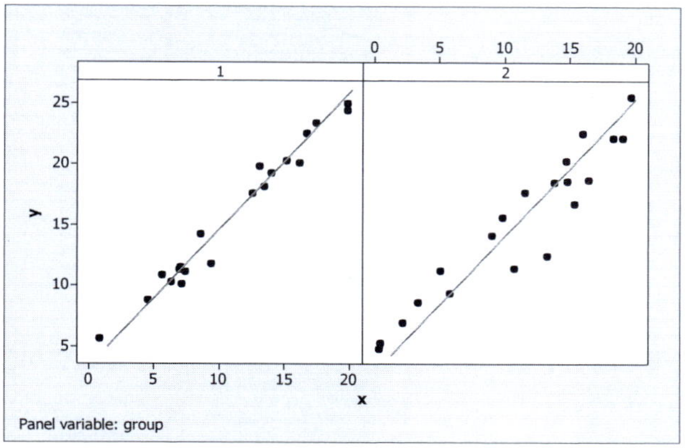

Figure 9.13: Two samples of (x, y) pairs with the same least squares line but different spreads around the fitted line

An additional measure is needed to describe how well the least-squares line fits a sample of bivariate data. The *correlation coefficient* is such a measure. The correlation coefficient for a sample of bivariate pairs (x, y) is defined as

$$r = \frac{\sum(x_i - \bar{x})(y_i - \bar{y})}{\sqrt{\sum(x_i - \bar{x})^2 \sum(y_i - \bar{y})^2}} \qquad (9.6)$$

It may be shown that r ranges from a minimum of -1 to a maximum value of 1. The correlation coefficient is a measure of the linear association in the sample between x and y.

For the two sets of data graphed in Figure 9.13, the correlation coefficient is $r = 0.987$ for the data on the left and $r = 0.949$ for the data on the right.

For the biscuit sales & price data, $r = -0.902$, while for the fuel consumption data r is 0.605, and for the trekker's weight loss the correlation coefficient is $r = 0.961$.

A correlation coefficient equal to 0 indicates no linear association between the two variables.

The *goodness-of-fit* of the least-squares line is defined as the square of the correlation coefficient (r^2).

The r^2 goodness-of-fit statistic may be expressed in terms of the sum of squares of the residuals and the sum of squares of the mean-corrected responses. If the fitted value of y corresponding to the determinant x_i is denoted by $\hat{y}_i$, so that

$$\hat{y}_i = a + b\,x_i \tag{9.7}$$

where a and b are given by Equations 9.4 and 9.5, then you can show that

$$r^2 = 1 - \frac{\sum(y_i - \hat{y}_i)^2}{\sum(y_i - \bar{y})^2} \tag{9.8}$$

Thus r^2 is the proportion of the squared variation in the responses accounted for by the fitted line.

For the biscuit sales-price data, the goodness-of-fit is 0.814, or 81.4%. The goodness-of-fit for the fuel consumption data is somewhat lower (36.6%), while for the trekker's weight loss, $r^2 = 0.923$.

For the data plotted in Figure 9.13, the goodness-of-fit is 0.974 for the data in the left-hand plot, compared with 0.901 for those in the right-hand plot.

Note that the goodness-of-fit of the fitted line ranges from 0 (when there is no correlation at all between x and y) to 1 or 100%, when there is perfect positive or negative correlation.

You cannot say what is an acceptable value for the goodness-of-fit statistic. It depends on the situation. But if you wish to give a verbal summary, you could say that the fit is 'poor' if r^2 is between 0 and 0.3, the fit is 'moderately good' if r^2 is between 0.3 and 0.7, 'very good' if r^2 is between 0.7 and 0.9, and 'excellent' if r^2 is greater than 0.9.

Confidence interval for β

Equation 9.1, that is,

$$y = \alpha + \beta x$$

specifies the *population* fitted line, and the parameter β in this equation is the population parameter corresponding to the slope b estimated from the sample (as the population mean μ is estimated by the sample mean $\bar{y}$). In this section you will learn how to obtain a confidence interval for β.

Note that an *error* may be defined as the difference between a response y_i and the predicted response using the population fitted line $\alpha + \beta x_i$, that is,

$$\varepsilon_i = y_i - (\alpha + \beta x_i) \tag{9.9}$$

These errors cannot be calculated because you don't know the population parameters α and β. However, you may use the sample estimates of the intercept a and the slope b to estimate the parameters α and β, so the residual $z_i = y_i - \hat{y}_i$, where $\hat{y}_i = a + b\,x_i$, may be used to estimate the error.

There is a simple formula for the standard error of b, namely

$$SE_b = \frac{\sigma}{\sqrt{\sum(x_i - \overline{x})^2}} = \frac{\sigma}{s_x\sqrt{n-1}} \tag{9.10}$$

where σ is the standard deviation of the distribution of the errors and s_x is the sample standard deviation of the x values. Now if σ were known, you could get a 95% confidence interval for β by taking an interval centered at b having width 3.92 times SE_b, based on the assumption that the responses are sampled from a normal distribution.

Since σ is usually unknown, the standard deviation s of the sample of residuals may be used to estimate it, based on the formula

$$s = \sqrt{\frac{\sum(y_i - \hat{y}_i)^2}{n-2}} \tag{9.11}$$

Note that Equation 9.11 is quite similar to the formula for the standard deviation of a sample of n observations. The difference is that, instead of subtracting the sample mean $\overline{y}$ to obtain residuals and dividing their sum of squares by $n-1$, you subtract the predicted value $\hat{y}_i$ from each response value and divide the sum of squares of these residuals by $n-2$. You divide by $n-2$ instead of $n-1$ because there are two parameters estimated in the line, rather than just one when estimating a mean.

So replacing σ by s in Equation 9.10, the estimated standard error of b is

$$s_b = \frac{s}{s_x\sqrt{n-1}} \tag{9.12}$$

Because of the additional uncertainty in not knowing σ, the normal distribution is no longer the appropriate distribution to use to obtain the confidence interval (unless the sample size is large). Instead, the *t* distribution should be used, and a 95% confidence interval for β is given by the formula

$$(\, b - t_{n\text{-}2}\, s_b \,,\;\; b + t_{n\text{-}2}\, s_b \,)\,, \tag{9.13}$$

where t_v is the critical value of Student's *t* distribution with ν degrees of freedom corresponding to a two-tailed area of 0.05.

Formula 9.11 is not designed for hand calculation, and the calculations are bothersome, requiring either tedious calculation of the predicted values and residuals, or some tricky short cut. It's best to store the data in a computer file and simply use a computer package to get the results. Most computer packages give the residual standard deviation *s*.

Consider the fuel consumption example. We find $b = 0.1162$, $s_x = 21.885$ and $s = 3.473$, so Equation 9.12 gives

$$s_b = \frac{3.473}{21.885\sqrt{15-1}} = 0.0424\,.$$

With 13 degrees of freedom, the critical value for *t* at the 5% level is (from Table A2) 2.160, so a 95% confidence interval for β is

$$(0.1162 - 2.160{\times}0.0424,\; 0.1162 + 2.160{\times}0.424) = \;(0.025,\, 0.208)$$

This means that you can say, with 95% confidence of being correct, that the slope of the line relating average speed to fuel consumption in the target population is somewhere between 0.025 and 0.208.

Testing the hypothesis that β = 0

There is an exact correspondence between a confidence interval for the slope parameter β and testing a hypothesis that β equals a given value. (This correspondence is due to fact that the standard error of *b* in Equation 9.12 does not depend on the population value β.)

If the specified value of β is 0, the null hypothesis states that there is no linear relation between *x* and *y*. The null hypothesis is thus

$$H_0{:}\;\; \beta = 0 \tag{9.14}$$

The alternative hypothesis is

$$H_1: \ \beta \neq 0$$

The test statistic for testing this null hypothesis against the alternative is

$$t = \frac{b}{s_b} \tag{9.15}$$

This statistic has Student's t distribution with $n-2$ degrees of freedom. A p-value may be obtained from Table A2.

For example, the t statistic for testing the nonexistence of a linear relation between average speed and fuel consumption is $t = 0.1162/0.0424 = 2.74$, and the number of degrees of freedom is $15-2 = 13$. From Table A2, the p-value corresponding to $t = 2.74$ and $\nu = 13$ is between 0.01 and 0.02. (A more accurate p-value is 0.0169.)

Assumptions

There are three assumptions underlying the statistical methods described in the preceding section, as follows:

a. The relation between x and y in the population is *linear*;

b. The standard deviation of the distribution of the errors is *constant*, that is, it does not depend on x;

c. The distribution of the errors is *normal*.

You may assess these assumptions by graphing residuals in two ways. The first graph involves plotting the residuals against the fitted values. Ideally, this plot will have no structure or pattern, since the residuals should have no trend if the relation is linear. Any curvature in the graph indicates a departure from the assumption of linearity.

Figure 9.14 shows plots of residuals versus fitted values for two of the data samples considered earlier in this chapter (the fuel consumption and the trekker's weight loss).

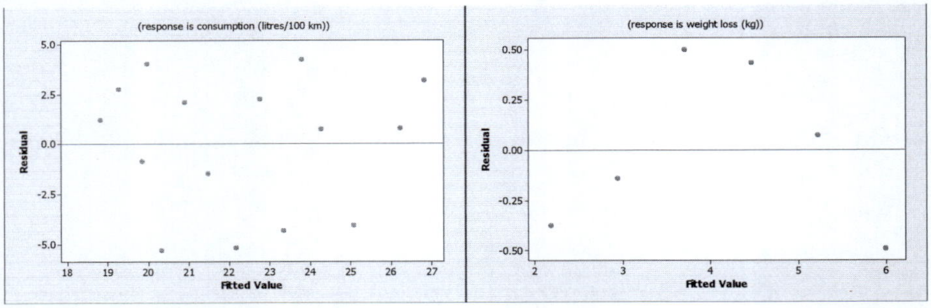

Figure 9.14: Plots of residuals versus fitted values

It is instructive to compare the left panel of Figure 9.14 with Figure 9.4: you can see that the points in the residuals plot are similarly situated in the two graphs, and no pattern is evident in these residuals.

Similarly, the right panel of Figure 9.14 should be compared with Figure 9.6. But the departure from linearity, or curvature, evident in Figure 9.6 is even more clearly evident in Figure 9.14.

Plotting residuals against fitted values is also a good way to check the second of the three assumptions required in linear regression analysis. If the errors do not have a constant standard deviation, the variation in the residuals will show a trend in the horizontal direction. Figure 9.15 gives an illustration of some data showing a tendency for the standard deviation of the residuals to increase with the fitted values, clearly demonstrating that the second of the assumptions is violated. These data comprise the rates (per 100,000) of burglaries and aggravated assaults in the 51 states of the USA (plus the District of Columbia) in 1986, taken from the Statistical Abstract of the U.S. (1988, page 265). The raw data are shown in Figure 9.16, which also includes other crime rates for the USA in the same year.

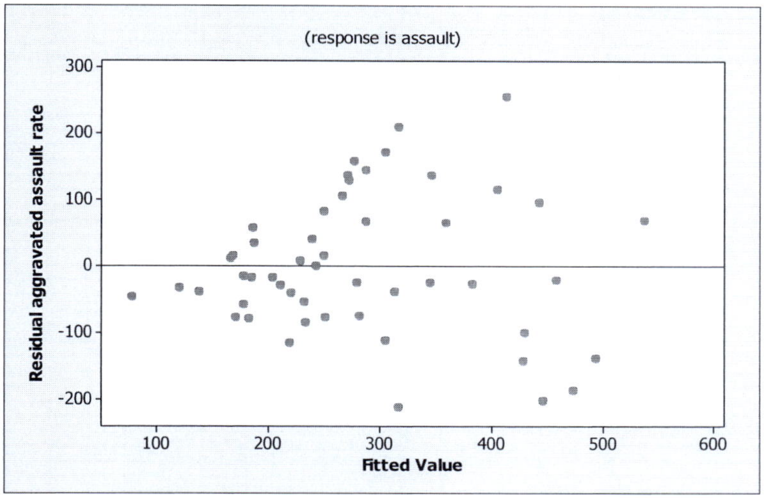

Figure 9.15: Plot of residuals versus fitted values showing a trend in the standard deviation of the errors

To assess the normality assumption it is best to plot the residuals against their corresponding normal scores, and you have already seen several examples of these plots.

When assessing relations it is especially important to ensure that outliers in the data do not unduly influence the fitted line. The data in Figure 9.17 were quoted in a letter to the *New York Times* on 25 November 1986. These comprise the death rates (per 100 million vehicle miles) and normal maximum speed limits (in miles per hour) for ten industrialised countries. Figure 9.18 shows a scatter plot of these data, with the fitted least-squares line shown as the continuous line.

State	*murder*	*rape*	*robbery*	*assault*	*burglary*	*larceny*	*mv theft*
Maine	2.0	14.8	28	102	803	2347	164
N.Hamp.	2.2	21.5	24	92	755	2208	228
Vermont	2.0	21.8	22	103	949	2697	181
Mass.	3.6	29.7	193	331	1071	2189	906
Rhode I.	3.5	21.4	119	192	1294	2568	705
Conn.	4.6	23.8	192	205	1198	2758	447
N.York	10.7	30.5	514	431	1221	2924	637
N.Jersey	5.2	33.2	269	265	1071	2822	776
Penn.	5.5	25.1	152	176	735	1654	354
Ohio	5.5	38.6	142	235	988	2574	376
Indiana	6.0	25.9	90	186	887	2333	328
Illinois	8.9	32.4	325	434	1180	2938	628
Michigan	11.3	67.4	301	424	1509	3378	800
Wisc.	3.1	20.1	73	162	783	2802	254
Minn.	2.5	31.8	102	148	1004	2785	288
Iowa	1.8	12.5	42	179	956	2801	158
Missouri	9.2	29.2	170	370	1136	2500	439
N.Dakota	1.0	11.6	7	32	385	2049	120
S.Dakota	4.0	17.7	16	87	554	1939	99
Nebraska	3.1	24.6	51	184	748	2677	168
Kansas	4.4	32.9	80	252	1188	3008	258
Delaware	4.9	56.9	124	241	1042	3090	272
Maryland	9.0	43.6	304	476	1296	2978	545
D.C.	31.0	52.4	754	668	1728	4131	975
Virginia	7.1	26.5	106	167	813	2522	219
W.Virg.	5.9	18.9	41	99	625	1358	169
N.Carol.	8.1	26.4	88	354	1225	2423	208
S.Carol.	8.6	41.3	99	525	1340	2846	277
Georgia	11.2	43.9	214	319	1453	2984	430
Florida	11.7	52.7	367	605	2221	4373	598
Kentucky	6.7	23.1	83	222	824	1740	193
Tenn.	10.4	47.0	208	274	1325	2126	544
Alabama	10.1	28.4	112	408	1159	2304	267
Miss.	11.2	25.8	65	172	1076	1845	150
Arkansas	8.1	28.9	80	278	1030	2305	195
Louisiana	12.8	40.1	224	482	1461	3417	442
Okla.	8.1	36.4	107	285	1787	3142	649
Texas	13.5	51.6	240	354	2049	3987	714
Montana	2.9	17.3	20	118	783	3314	215
Idaho	3.2	20.0	21	178	1003	2800	181
Wyoming	5.3	21.9	22	243	817	3078	169
Colorado	7.0	42.3	145	329	1792	4231	486
N.Mex.	11.5	46.9	130	538	1845	3712	343
Arizona	9.3	43.0	169	437	1908	4337	419
Utah	3.1	25.3	59	180	915	4074	223
Nevada	12.5	64.9	287	354	1604	3489	478
Wash.	5.0	53.4	135	244	1861	4267	315
Oregon	6.6	51.1	206	286	1967	4163	402
Calif.	11.3	44.9	343	521	1696	3384	762
Alaska	8.6	72.7	88	401	1162	3910	604
Hawaii	4.8	31.0	106	103	1339	3759	328

Figure 9.16: Crime rates (per 100,000 residents) in the US in 1986
Copyright US Department of Justice

Country	Death rate	Speed limit
Norway	3.0	55
United States	3.3	55
Finland	3.4	55
Britain	3.5	70
Denmark	4.1	55
Canada	4.3	60
Japan	4.7	55
Australia	4.9	60
Netherlands	5.1	60
Italy	6.1	75

Figure 9.17: Motor vehicle death rates (per100 million vehicle miles) and maximum speed limits in miles per hour for ten countries in 1986

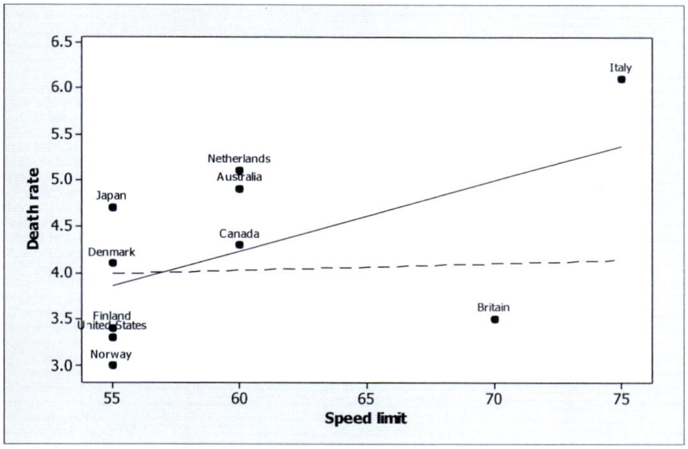

Figure 9.18: Relation between motor vehicle death & maximum speed limit

The fitted linear relation has a positive slope, suggesting that higher maximum speed limits are associated with more deaths. The correlation coefficient is 0.55. Note that the country with the highest maximum speed limit (Italy) also has the highest death rate. If this point is omitted from the plot, the relation disappears. The dotted line on the graph shows the fitted linear relation with Italy removed.

The lesson is that you should not believe a result that is dependent on a single data value. For these data, Italy is an outlier. Outliers are harmful when fitting a linear relation to a scatter plot, because they can distort the fitted line. Fortunately, it is easy to spot outliers in scatter plots.

Plotting residuals can be very informative, but the first thing to look at is the scatter plot itself. Often you can see just by looking at this graph whether a linear relation is appropriate or not.

Consider the data listed in Figure 9.19. These are the estimated surface temperatures and light intensities (expressed as base 10 logarithms) of 47 stars in the CYG OB1 cluster (see Rousseeuw & Leroy, 2005, page 27).

Index	Log surface temperature	Log light intensity	Index	Log surface temperature	Log light intensity
1	4.37	5.23	25	4.38	5.02
2	4.56	5.74	26	4.42	4.66
3	4.26	4.93	27	4.29	4.66
4	4.56	5.74	28	4.38	4.90
5	4.30	5.19	29	4.22	4.39
6	4.46	5.46	30	3.48	6.05
7	3.84	4.65	31	4.38	4.42
8	4.57	5.27	32	4.56	5.10
9	4.26	5.57	33	4.45	5.22
10	4.37	5.12	34	3.49	6.29
11	3.49	5.73	35	4.23	4.34
12	4.43	5.45	36	4.62	5.62
13	4.48	5.42	37	4.53	5.10
14	4.01	4.05	38	4.45	5.22
15	4.29	4.26	39	4.53	5.18
16	4.42	4.58	40	4.43	5.57
17	4.23	3.94	41	4.38	4.62
18	4.42	4.18	42	4.45	5.06
19	4.23	4.18	43	4.50	5.34
20	3.49	5.89	44	4.45	5.34
21	4.29	4.38	45	4.55	5.54
22	4.29	4.22	46	4.45	4.98
23	4.42	4.42	47	4.42	4.50
24	4.49	4.85			

Figure 9.19: Data from the CYG OB1 star cluster

These data are graphed as a scatter plot in Figure 9.20, with the fitted least squares line shown as the continuous line.

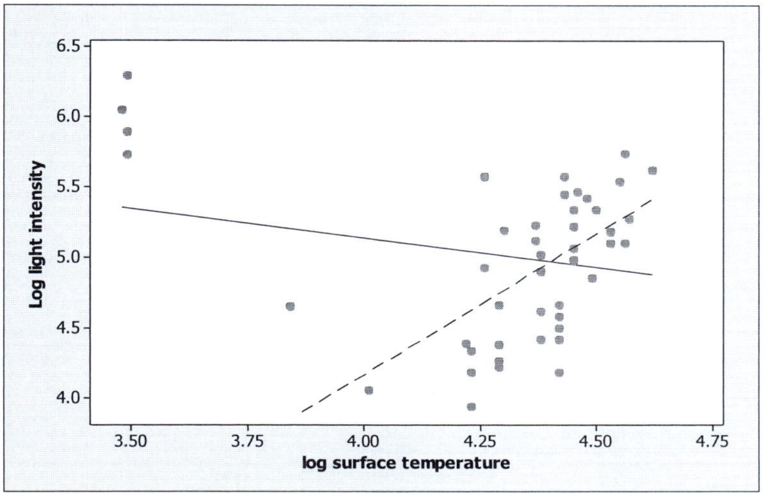

Figure 9.20: Scatter plot of CYG OB1 star cluster data with fitted lines

Clearly, the fitted line does not summarise the relation very well, for the obvious reason that the stars are clustered into two groups. The main group contains 43 stars and there is a smaller group containing four stars, which appear in the top left corner of the scatter plot.

The four stars appearing in the smaller cluster are actually *giant* stars. In comparison to the stars in the main group the giants have low surface temperatures and high light intensities. If these giants are omitted the fitted straight line (shown as the dotted line in Figure 9.20) is quite different, and is also quite easy to interpret. As you would expect from physics considerations, the higher the surface temperature, the higher the light intensity.

Summary

In this chapter you have seen methods for analysing bivariate data. The data thus comprise a sample of n pairs of the form (x, y), where x is a determinant and y is an outcome variable. It is assumed that both x and y are continuous variables.

To begin with, the data should first be displayed as a scatter plot, in which x is plotted on the horizontal axis and y on the vertical axis.

The relation between the two variables may be summarised by fitting a straight line to the points in the scatter plot. The least squares-fitted line, also called the regression line, is the one that minimises the sum of the squares of the distances of the points to the line in the vertical direction. This fitted line has the equation $y = a + b\,x$, where a and b are constants computed from the formulas

$$b = \frac{\sum(x_i - \bar{x})(y_i - \bar{y})}{\sum(x_i - \bar{x})^2} \quad \text{and} \quad a = \bar{y} - b\bar{x}$$

A measure of the strength of the linear relation between x and y is given by the correlation coefficient, defined as

$$r = \frac{\sum(x_i - \bar{x})(y_i - \bar{y})}{\sqrt{\sum(x_i - \bar{x})^2 \sum(y_i - \bar{y})^2}}$$

The value of r lies between -1 and 1. If r is 0, you can say that there is no linear relation in the sample, while if the magnitude of r is close to 1, there is a strong relation between the two variables in the sample.

The 'goodness-of-fit' of the fitted line is defined simply as the square of the correlation coefficient, r^2, and this may be interpreted as the proportion of the variation in the y values accounted for by the fitted line.

The fitted value of y corresponding to the determinant x_i is denoted by $\hat{y}_i$, and is expressed as $\hat{y}_i = a + b\,x_i$. The corresponding residual is $y - \hat{y}_i$. It is usually of interest to test the null hypothesis that $\beta = 0$, where β is the slope of the fitted line in the target population. To test this hypothesis, a t statistic (having n–2 degrees of freedom) may be used, namely

$$t = \frac{b}{s_b}$$

where s_b is the standard error of b, and may be computed from the formula

$$s_b = \frac{s}{s_x\sqrt{n-1}}$$

where s_x is the standard deviation of the sample of x values, and s is the standard deviation of the residuals. A p-value is obtained from the two-tailed area of the t distribution with n–2 degrees of freedom.

There are three assumptions required for the t test to be a valid procedure, namely (a) that the relation in the target population is linear, (b) that the errors from this relation in the y direction have constant standard deviation, and (c) that the distribution of these errors is normal. These assumptions may be assessed by plotting the residuals against the fitted values (to assess the first two assumptions) and by plotting the residuals against normal scores (to assess the normality assumption).

Finally, it is also important to look at a scatter plot of the data. Look for outliers that can distort the fitted relation, and also to see if there are other features (such as clustering of the data into two or more groups) that could complicate the relation between the two variables.

When examining linear relations in data, it is extremely useful to give a graphical summary of the analysis. This graph should contain a scatter plot of the data with the fitted line, the equation of the fitted line and the p-value for testing the hypothesis that the slope in the population is zero, and plots of residuals versus both the fitted values and the normal scores. Figure 9.22 shows such a graphical summary for a historical set of data of interest, listed in Figure 9.21. These are crew sizes and tonnages for 25 British merchant ships in 1907 (see Floud, 1973, Table 4.1).

Tonnage	*Crew size*	*Power*
44	3	not given
144	6	not given
150	5	not given
236	8	sail
739	16	steam
970	15	steam
2371	23	steam
309	5	steam
679	13	steam
26	4	sail
1272	19	steam
3246	33	steam
1904	19	steam
357	10	steam
1080	16	steam
1027	22	steam
45	2	not given
62	3	not given
28	2	sail
2507	22	steam
138	2	sail
502	18	steam
1501	21	steam
2750	24	steam
192	9	steam

Figure 9.21: Characteristics of 25 British merchant ships in 1907

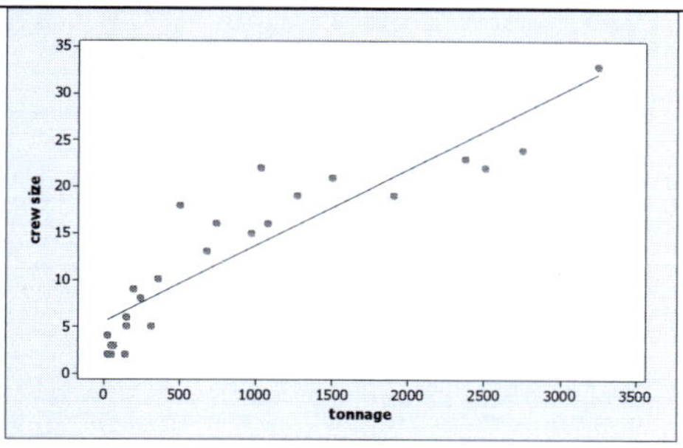

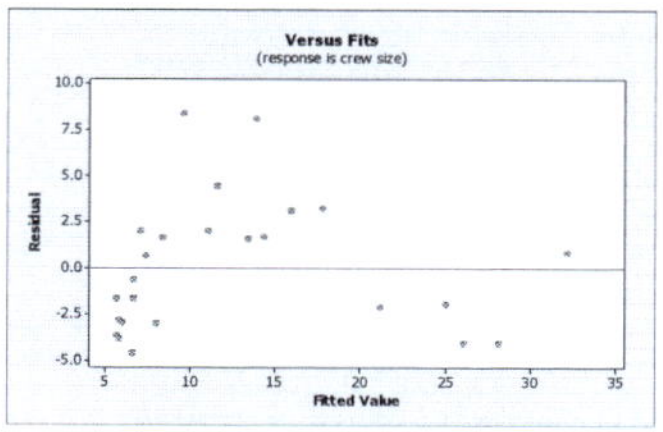

The regression equation is
crew size = 5.47 + 0.00823 tonnage

Predictor	Coef	SE Coef	T	P
Constant	5.466	1.016	5.38	0.000
tonnage	0.0082294	0.0007812	10.53	0.000

S = 3.70206 R-Sq = 82.8% R-Sq(adj) = 82.1%

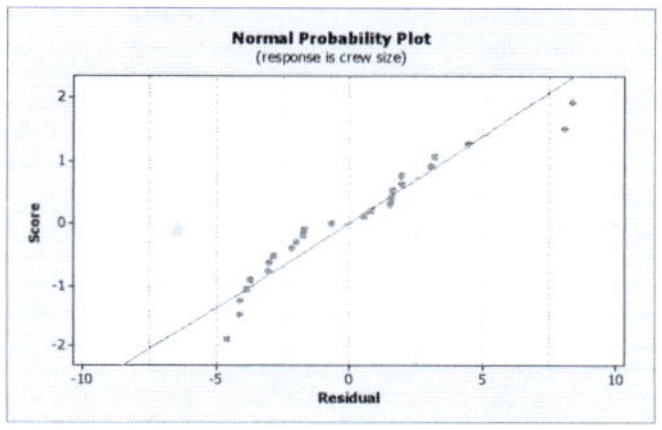

Figure 9.22: Graphical analysis of data for 25 British merchant ships

As you can see from the graph of residuals versus fitted values, the linearity assumption is suspect. Consequently the results are not valid.

It would be interesting to try to find a better fitting relation. One approach would be to transform the data before fitting a linear relation. This is left as an exercise for keen students.

Homework problems

1. (Taken from Utts, 1996, page 178) A professor sampled eight hardcover books from his shelf and recorded the number of pages and the price of each book, as follows:

pages	104	188	220	264	342	417	458	466
price	32.95	24.95	49.95	79.95	49.95	39.75	60.00	49.95

   Is there a relation between the number of pages and the price of a book? Answer this question by plotting the data as a scatter plot, fitting the regression line, and testing the appropriate null hypothesis. Interpret your findings. Use your fitted relation to predict the price of a hardcover book with 200 pages.

2. Age, gender, and smoking specific term life premium rates offered by the NRMA Insurance Company in 1985, for a sum insured of $50,000,are as follows (source: *The Open Road* magazine, June 85).

	Male		*Female*	
Age	*Smoker*	*Non-smoker*	*Smoker*	*Non-smoker*
34	135	105	110	95
36	145	110	120	100
38	160	115	130	105
40	180	125	145	115
42	210	140	165	130
44	250	155	190	145
46	295	180	225	165
48	360	215	265	195
50	435	260	320	230
52	535	315	380	275
54	650	390	460	335

   Choose one of these four categories (e.g. male non-smokers) and fit an appropriate linear model to predict premium from age. Does this model provide a reasonable fit to the data? Could you use the model to estimate the premium for a person aged 60? Explain.

3. The following data, compiled by UNICEF in 1992, show the under 5 mortality rate (per 1000 live births) and the adult female literacy rate (percent) for 11 Middle Eastern countries (with two missing literacy rates). Use these data from the nine countries having complete data to fit a linear model, and then attempt to use this model to estimate the missing data. Do you this think this approach is accurate?

Country	*child mortality*	*female literacy*
Yemen	187	21
Saudi Arabia	91	48
Iraq	86	49
Syria	59	51
Iran	59	43
Lebanon	56	73
Jordan	52	70
Oman	49	
U. Arab Emirates	30	38
Kuwait	19	67
Israel	11	

4. The ages of the subjects, whose body fat percentages are listed in Figure 2.1, are listed in Homework Problem 5 of Chapter 6. Investigate the relation between body fat percentage and age

   a. for the two sexes combined, and
   b. for females only.

   Do you think it is reasonable to combine the data for the two sexes when investigating this relation? Explain.

5. The following data (provided by Dr Eileen Gallery from Royal North Shore Hospital in Sydney) comprise the estimates of foetal weight (in grams) at successive visits by a pregnant woman to a clinic and the actual birth weight at 40 weeks. Fit a linear model to the data for the five visits, and compare the predicted value given by this model with the birth weight. Do you think it is possible, based on the model, to give a confidence interval for the birth weight?

week	20	24	30	34	36	**40**
weight	460	980	2000	2500	3000	**3175**

6. An experiment was conducted by Atwell (1991) to determine factors that affect the growth of grain legumes. Varieties of lupins and peas were grown in the field for 43 days on a solonised brown soil. During the course of the study the following air-filled porosity percentages were recorded at various depths below the soil surface.

	Time after sowing (days)			
Depth (cm)	22	29	36	43
5	23	25	21	17
12	19	17	20	15
20	15	14	24	16
28	17	14	24	15
36	31	13	16	20

a. Draw a scatter plot of the data relating air-filled porosity at 43 days to the porosity at 22 days, and comment on its appearance.

b. Compute the least-squares regression line relating the porosity at 43 days to that at 22 days, and give a measure of the strength of this relation.

c. Is knowing the porosity at 22 days useful for predicting it at 43 days? Perform an appropriate statistical test to answer this question.

d. Looking at the whole set of data, do you think that the depth below the surface affects the outcome? Give reasons for your answer.

e. Assuming that the depth below the surface does not affect the outcome, investigate the overall relation between the outcome and the time after sowing.

7. The Mooney Problem Checklist (MPCL) comprises 330 items classified into 11 problem areas, and highlights the nature of personal problems and their relation to self-esteem. In a study by Harper & Marshall (1991), the MPCL was administered to 201 adolescents (101 girls and 100 boys) from four secondary co-educational schools in the Sydney metropolitan area, and the following average scores were obtained.

   a. Compute the correlation coefficient between the scores for the girls and those for the boys.

   b. Compute the goodness-of-fit of the relation. Describe how well the linear relation fits the data.

   c. Display the data in a scatter plot, with the girls' scores plotted on the horizontal axis.

   d. Fit the least-squares line to the scatter plot.

   e. If you had plotted the boys' scores on the horizontal axis and fitted the least-squares line to this scatter plot, would you get the same fitted relation? Explain. If not, discuss which line is more appropriate?

MPCL Problem Area	*girls*	*boys*
Health & physical development	4.27	2.83
Finance, living conditions, employment	4.72	3.56
Social & recreational activities	3.78	3.12
Courtship, sex & marriage	4.57	3.27
Social psychological relationships	5.81	4.01
Personal psychological relationships	5.09	3.35
Morals & religion	3.89	3.07
Home & family	4.09	2.95
The future, vocational & employment	3.54	2.93
Adjustment to school work	5.60	5.70
Curriculum & teaching procedures	4.52	4.66

8. Dr M. Hills from the Open University in England collected the following chest, waist and hips measurements, in inches, for 10 men and 10 women (cited on page 62 of Hand et al, 1994). (The men and the women are not identified.)

Individual	*Chest*	*Waist*	*Hips*
1	34	30	32
2	37	32	37
3	38	30	36
4	36	33	39
5	38	29	33
6	43	32	38
7	40	33	42
8	38	30	40
9	40	30	37
10	41	32	39
11	36	24	35
12	36	25	37
13	34	24	37
14	33	22	34
15	36	26	38
16	37	26	37
17	34	25	38
18	36	26	37
19	38	28	40
20	35	23	35

If you know the sum of the waist and hips measurements for a person, how well can you predict their chest measurement? Answer this question with the given data by doing an appropriate linear regression analysis.

9. Using the data from Figure 9.16, investigate the relation between the murder rate and the motor vehicle theft rate. Choose the murder rate as the outcome variable. Should the value for the District of Columbia be omitted before fitting the regression line? Explain.

10. Gilchrist (1984, page 5) gives the following distances by road and the corresponding distances as the crow flies (each measured in miles) between 20 pairs of points in Sheffield, England. What is the relationship between the two variables? How well can the road distance be predicted from the distance as the crow flies?

Road distance	Direct distance	Road distance	Direct distance
10.7	9.5	11.7	9.8
6.5	5.0	25.6	19.0
29.4	23.0	16.3	14.6
17.2	15.2	9.5	8.3
18.4	11.4	28.8	21.6
19.7	11.8	31.2	26.5
16.6	12.1	6.5	4.8
29.0	22.0	25.7	21.7
40.5	28.2	26.5	18.0
14.2	12.1	33.1	28.0

CROSS-TABULATIONS

CHAPTER 10

10 Cross-tabulations

In the preceding chapter you saw examples of bivariate data involving a continuous outcome and a continuous determinant. In these situations the question of interest usually involves the association between the two variables. This association is called a relation, and may be summarised by a fitted line.

In this chapter we consider examples in which both the determinant and the outcome are categorical. In this case the association may be shown in a contingency table, which is a cross-tabulation of the two variables.

In the very simplest case when both the outcome and the determinant are binary variables, the data may be presented as a table with four cells (called a 2×2 contingency table). The cells in this table contain the numbers of observations corresponding to each combination of determinant and outcome. The question of interest involves comparing the proportions having a given outcome in the two determinant groups. In Chapter 8 you saw how a z statistic could be used to obtain a p-value for testing the null hypothesis that these population proportions are equal, based on the normal distribution. Recall that different formulas are needed, depending on whether the samples are independent or matched.

The difference between two proportions may be used as a measure of the association between a binary determinant and a binary outcome. In this chapter you will meet another important measure of this association called the odds ratio.

This chapter also describes more general methods for analysing data in which both the determinant and the outcome are categorical. Such data may be summarised by a table with several rows and several columns (called an $r \times c$ contingency table, r being the number of categories defining the determinant and c the number of categories in the outcome).

Before examining associations between determinants and outcomes, we begin with the even simpler situation corresponding to a single sample of categorical data. In this case it is of interest to compare the observed proportions in the various outcome categories with some hypothesised distribution, giving rise to what is known as a *goodness-of-fit* test. If the outcome has just two categories, the procedure we use here is essentially the same as the method given in Chapter 7 for testing that a population proportion has a given value (Equation 7.5).

Testing goodness-of-fit

Categorical data are nominal rather than numerical. Questionnaires used in surveys ask respondents to tick boxes to select categories most closely matching their situations or opinions. For example, age groups or income ranges or educational levels are often given, from which the respondent is asked to select the appropriate category.

Counting the number of individuals selecting each response category tallies categorical data. These categories must not overlap and each piece of information contributes to one and only one category.

Consider the preferences of 90 conference attendees for tea, coffee, or orange juice during the morning refreshment break. The conference organisers would like to know how many beverages of each type to supply. Short of doing some research to find out the distribution of preferences of their attendees, they may be tempted to assume that the proportions are equal. In statistical terms, this corresponds to the null and alternative hypotheses

$$H_0: \pi_1 = \pi_2 = \pi_3$$

$$H_0: \text{ the proportions are not equal}$$

Recall that the general method for statistical hypothesis testing is to first state the null and alternative hypotheses, and then to collect some data and compare the data with the null hypothesis, obtaining a p-value as a measure of the plausibility of the null hypothesis. The null hypothesis should not be too complicated, and conventionally translates as 'no difference' or 'no association', or 'the status quo'. The above null hypothesis, which reflects no difference between the three available drink preferences, follows this convention.

Suppose that 20 persons choose tea, 40 coffee and 30 juice. Are these data consistent with the null hypothesis? How do you get a p-value?

We need to use some method to compare this pattern of preferences observed in the sample (and our conference group is just one sample) with the pattern we would expect to see if the null hypothesis was true. Clearly, if there was no difference in the preferences for the three beverages, we would expect that, of the 90 persons at the conference, about 30 would drink tea, 30 coffee, and 30 orange juice. These expected counts could be compared with the 20 tea, 40 coffee, and 30 orange juice drinkers observed in the sample. The pattern seems to be different but, as you know from other statistical tests, we need to establish whether it's likely that the discrepancy is attributable to sampling variability. If not, the differences can be judged to be statistically significant, and the null hypothesis of preference of equality is rejected.

The conventional method for approaching this problem involves the *chi-squared goodness-of-fit test*. This procedure may be described in general as follows.

The null hypothesis may be stated as

$$H_0 : \pi_1 = \pi_{10} , \pi_2 = \pi_{20} , \ldots , \pi_c = \pi_{c0} \qquad (10.1)$$

where c is the number of categories for the outcome variable. (Note that there are c statements here, but only $c-1$ statements are needed because the proportions sum to 1.)

The test statistic takes the form

$$\chi^2 = \sum_{j=1}^{c} \frac{(O_j - E_j)^2}{E_j} \qquad (10.2)$$

where O_j is the observed count in category j and E_j is the expected count, assuming that the null hypothesis is true. The expected count in category j is obtained simply by multiplying the proportion specified in the null hypothesis by the sample size n, that is

$$E_j = n\pi_j \qquad (10.3)$$

The chi-squared distribution is used to obtain a p-value, in the same way as the t distribution gives p-values for the t tests described in preceding chapters. Like the t distribution, the chi-squared distribution has parameter ν corresponding to a number of degrees of freedom. For the chi-squared goodness-of-fit test, this parameter is one less than the number of categories, that is, $\nu = c-1$.

The chi-squared distribution differs from the t distribution in one important respect: it is not symmetric, and takes only nonnegative values. As a consequence, the p-value is defined as the area in the *upper* tail of the distribution, rather than the area in both tails. This corresponds to the fact that a goodness-of-fit test should give a statistically significant result only if the discrepancy between the observed and expected cell counts is large. A graph of a chi-squared distribution is shown in Figure 10.1 (in the case when ν is 5), and Table A3 in Appendix A gives chi-squared values corresponding to specified p-values in the same way as the t-table.

Now let us consider the application of the chi-squared test to the drink preferences. The null hypothesis that the proportions are the same may be restated in the form of Equation 10.1 as

$$H_0: \pi_1 = \tfrac{1}{3}, \pi_2 = \tfrac{1}{3}, \pi_3 = \tfrac{1}{3}$$

and the alternative hypothesis as

$$H_1 \text{ the proportions are not equal}$$

Note that the observed counts are $O_1 = 20$, $O_2 = 40$, and $O_3 = 30$. The sample size is $n = 90$, so Equation 10.3 gives $E_1 = E_2 = E_3 = 90 \times \tfrac{1}{3} = 30$. Substituting these quantities in Equation 10.2, we find

$$\chi^2 = \frac{(40-30)^2}{30} + \frac{(20-30)^2}{30} + \frac{(30-30)^2}{30}$$
$$= 3.333 + 3.333 + 0 = 6.666$$

The number of degrees of freedom is 3−1 = 2. From Table A3, the p-value is between 0.02 and 0.05. (The exact p-value given by a statistical package is 0.036.)

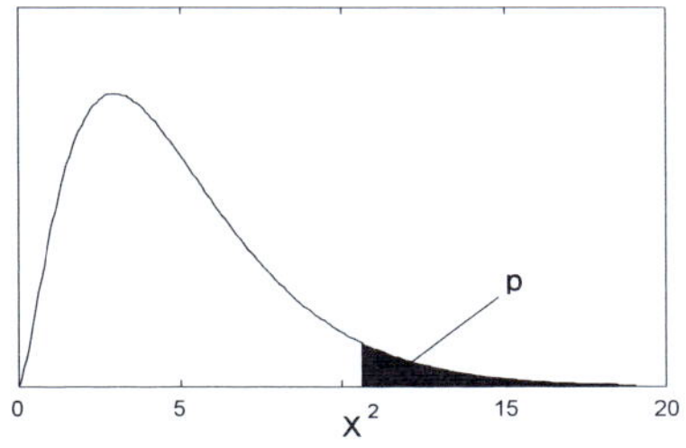

Figure 10.1: Tail area of a chi-squared distribution with ν = 5

Given that a small p-value is obtained, the null hypothesis is rejected. Conclude that the drink preferences are not in equal proportions.

Note that if there are just two categories, the null hypothesis reduces to

$$H_0: \pi_1 = \pi_{10}, \pi_2 = \pi_{20}$$

And, since the two population proportions add up to 1, this simplifies to

$$H_0: \pi_1 = \pi_{10}$$

This is essentially the same as the null hypothesis given in Chapter 7 (Equation 7.5) where you saw that a z statistic (given by Equation 7.6) could be used to obtain a p-value.

In fact statistical theory may be used to show that the z test based on Equation 7.2 gives exactly the same p-value as the chi-squared goodness-of-fit test based on Equation 10.2. In this case you can show that the chi-squared statistic given by Equation 10.2 (for which the number of degrees of freedom is 1) is equal to the square of the z statistic given by Equation 7.5. And a chi-squared statistic with $\nu = 1$ has exactly the same distribution as the square of a z statistic.

In the next section we consider the case when there are two or more categories for the determinant. In this situation the question of interest is to compare the pattern of outcomes corresponding to the different categories in the determinant. The method again involves a chi-squared test.

Exercise 10.1:

There were 81 claimants in a sample of 400 car insurers. Apply the chi-squared statistic to test the hypothesis that the population proportion of claimants is 0.25. Compare your answer with that based on Equation 7.2.

Testing independence

Consider the following data from a study, reported by Brown and Fielding (1993), to investigate demographic factors and their effects on occupational stress among officers in a large provincial British police department. The association between gender and education is shown in Figure 10.2.

gender	*no formal qualifications*	*school*	*university*	*other*
women	3	91	11	5
men	23	231	19	39

Figure 10.2: Education by gender for 422 British police officers

The outcome variable (level of education) has four categories and the determinant (gender) is binary. The question of interest is 'do women police officers have different levels of education than male colleagues?'

Now of course there are fewer women than men in the study, undoubtedly reflecting the fact that women are less inclined than are men to choose a career in the police force. So we need to examine the proportions rather than the counts: these are given in Figure 10.3.

gender	*no formal qualifications*	*school*	*university*	*other*
women	2.7	82.7	10.0	4.5
men	7.1	71.7	6.1	12.1

Figure 10.3: Gender and education association for 422 British police officers, expressed as percentages

(These proportions are obtained by dividing the numbers of women at each educational level by 110, the total number of women in the study, and doing an equivalent calculation for the men. The percentages in each column don't quite add up to 100 due to rounding error.)

Women have more than their share of school and university graduates but substantially less than their share of those with other tertiary qualifications. However there is a greater proportion of men with no formal qualifications.

Before jumping to conclusions, a statistical test is needed. The appropriate null hypothesis may be written as

$$H_0 : \pi_{11} = \pi_{21} , \pi_{12} = \pi_{22} , \pi_{13} = \pi_{23} , \pi_{14} = \pi_{24}$$

where π_{1j} and π_{2j} are the population proportions of women and men, respectively, having educational level in category *j*. (For definiteness, the education categories are coded as 1 for no formal qualifications, 2 for school only, 3 for university, and 4 or other qualification.)

The null and alternative hypotheses can be more simply stated as

H_0 : there is no association between gender and educational level

H_0 : there is an association between gender and educational level

The procedure for testing this null hypothesis is similar to that for the goodness-of-fit test described in the preceding section. First, compute a chi-squared statistic, and then get a p-value based on the chi-squared distribution with a specified number of degrees of freedom. The chi-squared statistic is again based on the comparison between the observed counts and expected counts based on the null hypothesis and is defined in general terms as follows.

The null hypothesis has the general form

$$H_0 : \pi_{11} = \pi_{21} , \pi_{12} = \pi_{22} , \dots , \pi_{1c} = \pi_{2c} \qquad (10.4)$$

The test statistic is defined as

$$\chi^2 = \sum_{j=1}^{c} \frac{(O_{1j} - E_{1j})^2}{E_{1j}} + \sum_{j=1}^{c} \frac{(O_{2j} - E_{2j})^2}{E_{2j}} \qquad (10.5)$$

As in the case of Equation 10.2, c is the number of categories for the outcome variable, and the symbols O and E refer to observed and expected counts in these categories. Because the determinant is binary and thus has two categories, the chi-squared statistic comprises two components, each of which is a sum of contributions from each outcome category. Thus O_{1j} is the observed count in outcome category j and the first category of the determinant, and similarly O_{2j} is the observed count in outcome category j and the second category of the determinant.

The corresponding expected counts are calculated by apportioning the total number of observations in each outcome category to the two determinant categories according to their total numbers of outcomes

$$E_{1j} = \frac{(O_{1j} + O_{2j})\sum_{j=1}^{c} O_{1j}}{n}, \quad E_{2j} = \frac{(O_{1j} + O_{2j})\sum_{j=1}^{c} O_{2j}}{n} \qquad (10.6)$$

where n is the total sample size. The number of degrees of freedom associated with the chi-squared statistic is $c-1$.

Although they look complicated, these formulas are really quite simple. Consider the data from Figure 10.4 giving the observed cross-tabulation by gender and educational level for the British police officers. Table 10.4 is the same as Figure 10.3, with additional row and column the totals.

gender	*no formal qualifications*	*school*	*university*	*other*	total
women	3	91	11	5	110
men	23	231	19	39	312
total	26	322	30	44	422

Figure 10.4: Observed counts with totals for gender and education association in study of British police officers

The expected counts are obtained simply by multiplying the marginal totals together and dividing the resulting products by the overall total. Thus E_{11} is the expected count for the women in the first outcome category ('no formal qualifications'), and is obtained by multiplying the marginal total for 'no formal qualifications' (26) by the marginal total for women (110) and dividing by 422: this gives $E_{11} = 26{\times}110/422 = 6.78$.

Applying this procedure to compute all the other expected counts, the following results are obtained:

$$E_{21} = 26{\times}312/422 = 19.22,$$

$$E_{12} = 322{\times}110/422 = 83.93, \quad E_{22} = 322{\times}312/422 = 238.07,$$

$$E_{13} = 30{\times}110/422 = 7.82, \quad E_{23} = 30{\times}312/422 = 22.18,$$

$$E_{14} = 44{\times}110/422 = 11.47, \quad E_{24} = 44{\times}312/422 = 32.53.$$

It remains to substitute these values into Equation 10.5 to get the chi-squared statistic. So we obtain

$$\chi^2 = \frac{(3-6.78)^2}{6.78} + \frac{(91-83.93)^2}{83.93} + \frac{(11-7.82)^2}{7.82} + \frac{(5-11.47)^2}{11.47}$$

$$+ \frac{(23-19.32)^2}{19.32} + \frac{(231-238.07)^2}{238.07} + \frac{(19-22.18)^2}{22.18} + \frac{(39-32.53)^2}{32.53}$$

$$= 2.107 + 0.596 + 1.293 + 3.650 + 0.743 + 0.210 + 0.456 + 1.287$$

$$= 10.342.$$

The number of degrees of freedom is $4-1 = 3$, so a p-value may be obtained by comparing 10.342 with the chi-squared table with $\nu = 3$. Table A3 thus gives a p-value between 0.01 and 0.02. (The exact p-value is 0.0159).

Since the p-value is small the null hypothesis is rejected, and you conclude that the men and women officers have different educational levels.

Note that you cannot say that the women are less well educated than the men. Nor can you say that the women are better educated than the men. You could draw such a conclusion if the differences were all in the same direction, that is, if the proportions of one group compared to the other changed in a consistent way as educational level increased. But the data do not show such a simple trend.

However, you can tell where the differences lie by examining the individual components of the chi-squared statistic. The largest values of these components correspond to the greatest discrepancies between the proportions in the two groups.

In this case the largest contributions to the chi-squared statistic correspond to the women in the two extreme educational categories: women with 'other' qualifications contribute 3.650 while women with no formal qualifications contribute 2.107. In each case there are fewer women in these categories than would be expected if the null hypothesis held.

If there are just two categories for the outcome as well as the determinant (that is, if both the determinant and the outcome are binary variables), c is 2, and the null hypothesis given by Equation 10.4 reduces to

$$H_0 : \pi_{11} = \pi_{21} , \pi_{12} = \pi_{22}$$

In fact, since the proportions in the two outcome categories for each level of the determinant add up to 1, (that is, $\pi_{11} + \pi_{21} = 1$ and $\pi_{12} + \pi_{22} = 1$), the second of the two equations in this null hypothesis is redundant, so it simplifies even further to

$$H_0 : \pi_{11} = \pi_{21}$$

which is essentially the same as the null hypothesis for comparing two proportions given as Equation 8.8.

The alternative hypothesis will be

$$H_1 : \pi_{11} \neq \pi_{21}$$

Moreover, it may be shown that the chi-squared test described above is precisely equivalent to the z test for comparing two population proportions defined in Equations 8.9 and 8.10.

Exercise 10.2: *Recall the study of knee injuries among skiers in the Snowy Mountains cited in Chapter 8, where it was found that 40 out of 197 men and 75 out of 308 women reported serious injuries. Use the chi-squared test to obtain a p-value for testing the null hypothesis that men and women skiers reporting knee injuries have the same proportions of serious injuries, and compare this p-value with that obtained from the z test.*

The method for comparing the population proportions in two groups classified by a dichotomous determinant extends in a straightforward way to situations involving more than two determinant categories. In this case the chi-squared statistic takes the form

$$\chi^2 = \sum_{i=1}^{r} \sum_{j=1}^{c} \frac{(O_{ij} - E_{ij})^2}{E_{ij}} \qquad (10.7)$$

where O_{ij} is the observed count in category i of the determinant and category j of the outcome, and E_{ij} is the corresponding expected count, defined as before by dividing the product of the marginal totals by the overall total sample size, that is

$$E_{ij} = \frac{\sum_{j=1}^{c} O_{ij} \sum_{i=1}^{r} O_{ij}}{n} \qquad (10.8)$$

The number of degrees of freedom is $\nu = (r-1)(c-1)$.

This test is called the chi-squared test for *independence*, because the null hypothesis states that the determinant and outcome variables are independent of each other. This is the same as saying that the determinant is not associated with the outcome, because if there is no association between them, the counts in the various categories for each variable arise independently of the other variable.

Exercise 10.3*: The following data were obtained from 413 patients in an experiment to investigate the association between ocularity and dexterity. Is there evidence of a connection between these two factors?*

	left-eyed	*ambiocular*	*right-eyed*	total
left-handed	34	62	28	124
ambidextrous	27	28	20	75
right-handed	57	105	52	214
total	118	195	100	413

The odds ratio

Another result obtained from the study of the British police officers (described in the preceding section) is that 51 of the 110 women officers (46.4%) described their marital status as single, compared with only 36 of the 312 men officers (11.5%). You don't need to be a statistical wizard to conclude that there is a statistically significant difference between these proportions. If you go through the calculations for the two-sample proportions test, you will find that Equation 8.7 gives a p-value of 0. You get exactly the same result if you use the chi-squared test for independence described in the preceding section.

The problem with this result is that it doesn't really tell you very much. To say that the p-value is 0, and therefore the proportions of single police officers are different for men and women, isn't very informative. What is needed is a *measure* of the difference, and preferably a confidence interval giving an indication of the probable range of this difference.

The difference between the proportions is one such measure, and it is possible to obtain a confidence interval by a simple extension of the method given in Chapter 6.

Another measure is the *odds ratio.* To get an understanding of an odds ratio, suppose the data are presented in a 2×2 contingency table as follows.

		marital status		
		single	*with partner*	total
gender	*men*	36	276	312
	women	51	59	110
	total	87	335	422

Figure 10.5 Marital status and gender of 422 British police officers

Now the *odds* of an event or outcome is defined as the ratio of the probability of the event to the probability of the corresponding non-event. Here we are using the term 'probability' to mean the proportion of outcomes in the population of interest. Thus if an event has probability π and the corresponding non-event has probability $1-\pi$ (so that there are only two possible outcomes) the odds of the event is $\pi/(1-\pi)$.

The estimated odds of an outcome is obtained by replacing the population proportion π by the sample proportion p, giving the formula $p/(1-p)$.

Applying this definition to the data shown in Table 10.5 where the sample proportion (p) of single police officers is 87/422, the estimated odds that a police officer is single is

$$\frac{p}{1-p} = \frac{87/422}{1-87/422} = \frac{87}{335} = 0.260.$$

The estimated odds of an outcome is thus the ratio of the number of outcomes of interest to the number of non-outcomes.

The odds is always greater than the corresponding proportion or probability. But unlike a proportion, which is restricted to being between 0 and 1, the odds can be greater than 1. (For example, if a proportion is 0.6 the corresponding odds is 0.6/(1–0.6) = 1.5.)

The *odds ratio* is associated with two proportions, and is simply defined as the ratio of the two odds. If the population proportions are π_1 and π_2, respectively, the odds ratio is

$$\omega = \frac{\pi_1/(1-\pi_1)}{\pi_2/(1-\pi_2)} \tag{10.9}$$

Similarly, the estimated odds ratio from a sample of data is obtained by replacing the population proportions π_1 and π_2 by their corresponding sample proportions p_1 and p_2. This gives

$$w = \frac{p_1/(1-p_1)}{p_2/(1-p_2)} \tag{10.10}$$

Now using the data in Figure 10.6, the estimated odds that a male officer is single is 36/276, while the estimated odds that a female officer is single is 51/59. So the estimated odds ratio of male to female officers being single is

$$w = \frac{36/276}{51/59} = 0.151.$$

The odds ratio is a measure of the strength of an association between two binary variables. If the two proportions being compared are equal, the odds ratio is 1, indicating that there is no association between the determinant and the outcome variables. But if the proportions are different from each other, the odds ratio is different from 1. It is less than 1 if π_1 is less than π_2, while if π_1 is greater than π_2 the odds ratio is greater than 1. For these data the estimated odds ratio for the outcome 'single' for men compared to women police officers is relatively small, indicating that women police officers are much more likely to be single than their male colleagues.

To illustrate further, let us have another look at the data involving men and women skiers referred to in Chapter 8.

The data may be presented in a contingency table as shown below in Figure 10.6.

	severe	*not severe*	total
men	40	157	197
women	75	233	308
total	115	390	505

Figure 10.6: Knee injuries of 505 skiers, classified by gender and severity

The estimated odds ratio of a severe injury for men compared to women is thus (40/157) / (75/233) = 0.792. Again this result is more informative than just saying, as we did in Chapter 8, that there is insufficient evidence of any difference in the risk of a serious injury for men and women skiers reporting injuries.

An even more informative result is obtained by adding a confidence interval, which we do in the next section.

Confidence interval for an odds ratio

A 2 ×2 contingency table has the following general form.

	outcome=1	*outcome=2*	total
determinant=1	a	c	$a+c$
determinant=2	b	d	$b+d$
total	$a+b$	$c+d$	$a+b+c+d$

Figure 10.7: General form of 2× 2 contingency table

In this notation, a and c are the numbers of outcomes coded as 1 and 2, respectively, in the first group (that is, where the determinant is coded as 1), while b and d are the numbers of respective outcomes in the second group. Thus Equation 10.10 may be expressed in terms of these counts as

$$w = \frac{a / c}{b / d} = \frac{a \times d}{b \times c} \qquad (10.11)$$

A confidence interval for the population odds ratio ω is based on the fact that the statistic

$$z = \frac{\ln(w) - \ln(\omega)}{s_w} \qquad (10.12)$$

where

$$s_w = \sqrt{\frac{1}{a} + \frac{1}{b} + \frac{1}{c} + \frac{1}{d}} \qquad (10.13)$$

is approximately normally distributed with mean 0 and standard deviation 1. (The logarithm in Equation 10.12 is the natural or base e logarithm.) It follows that a 95% confidence interval for ω is given by the formula

$$(\, w \exp(-1.96\, s_w),\; w \exp(1.96\, s_w)\,) \qquad (10.14)$$

Now let us use these formulas to get a confidence interval for the odds ratio for single marital status for men compared to women, based on the data in Figure 10.6. We have a = 36, b = 51, c = 276 and d = 59, so Equations 10.11 and 10.13 give

$$w = \frac{36 \times 59}{51 \times 276} = 0.151$$

(as before) and

$$s_w = \sqrt{\frac{1}{36} + \frac{1}{51} + \frac{1}{276} + \frac{1}{59}} = 0.261.$$

Substituting these values into Equation 10.14 gives the 95% confidence interval

$$(\, 0.151 \exp(-1.96 \times 0.261),\; 0.151 \exp(1.96 \times 0.261)\,) = (0.0905,\; 0.252).$$

So you may conclude that the population odds ratio could be as low as 0.0905 or as high as 0.252. This is a useful result, because it gives an upper bound to the likely odds ratio.

Graphing odds ratios

When measuring an association between two binary categorical variables, it is informative to graph the odds ratio, together with a 95% confidence interval based on Equation 10.14. Figure 10.8 shows such a graph for the data given in Figure 10.5. Note that the odds ratio is scaled in terms of powers of 2, so that each unit increase on the horizontal axis corresponds to a doubling of the odds ratio. The graph thus shows that the odds ratio could be as high as $2^{-2} = ¼$, and that the p-value for testing the null hypothesis that the odds ratio is 1 (represented by the dotted line) is extremely small.

Figure 10.9 shows a similar graph summarising the results of the study cited in Exercise 8.2. This graph is again very informative because it clearly shows that the study was inconclusive, even though the null hypothesis could not be rejected because of the relatively high p-value. The confidence interval shows that the odds ratio could be as low as 0.6 or as high as = 2.7. This means that at one extreme there could be a 40% reduction in mortality due to home care compared to hospital treatment, or at the other extreme the risk of death could be more than doubled. A much larger study would be needed to properly compare these alternative treatment strategies.

If there are more than two outcome categories it is still possible to calculate odds ratios. For each outcome category, you simply add up all the outcomes from the other 'non-outcome' categories, and separately compute an odds ratio for each cell in the table.

To illustrate this method, consider the data in Figure 10.4. If you select the category 'no formal qualifications' as the referent category, the odds ratios may be graphed as in Figure 10.10.

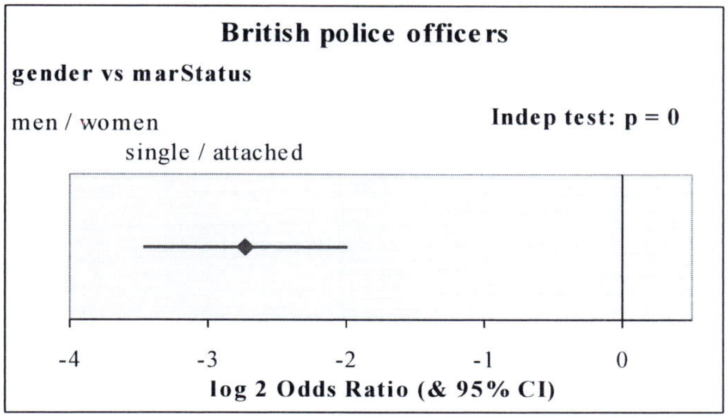

Figure 10.8: Odds ratio and 95% confidence interval for association between gender and marital status for 422 British police officers

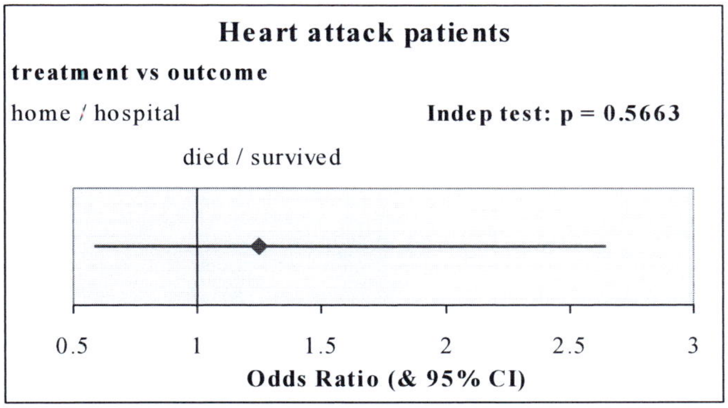

Figure 10.9 Odds ratio plot from randomised trial comparing home and hospital treatment for 264 heart attack patients

This graph shows that women police officers are less likely to be in the 'other' educational level category, and there is also a trend (not quite statistically significant, according to the confidence interval) for the women to be less likely than the men to have 'no formal qualifications'. In other words, there are disproportionately fewer women (or disproportionately more men) in the study having (a) no formal qualifications and (b) qualifications other than obtained from school or university.

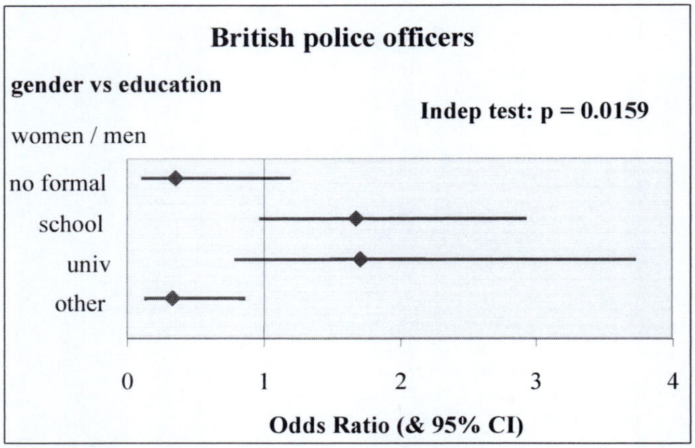

Figure 10.10: Odds ratios obtained by comparing educational levels of 422 men and women police officers in Britain.

What are the implications of these results? You would need to know more about the nature of provincial police forces in Britain before making any definite statements, but it could be that there is discrimination in the British police force against women without school or university qualifications.

Odds ratio plots may be generalised further to handle categorical data with several categories for the determinant as well as the outcome variable. The odds ratios are defined separately for each combination of determinant and outcome category, where the counts in the other categories are combined.

Consider the study of ocularity versus dexterity given in Exercise 10.3. It is not obvious which variable is the determinant and which is the outcome here (in fact both are outcomes influenced largely by genetics), but for definiteness let us choose dexterity as the determinant and ocularity as the outcome variable.

The odds ratios are graphed in Figure 10.11. The chi-squared test for independence does not give a significant result, and there is no evidence in the graph of any association between ocularity and dexterity.

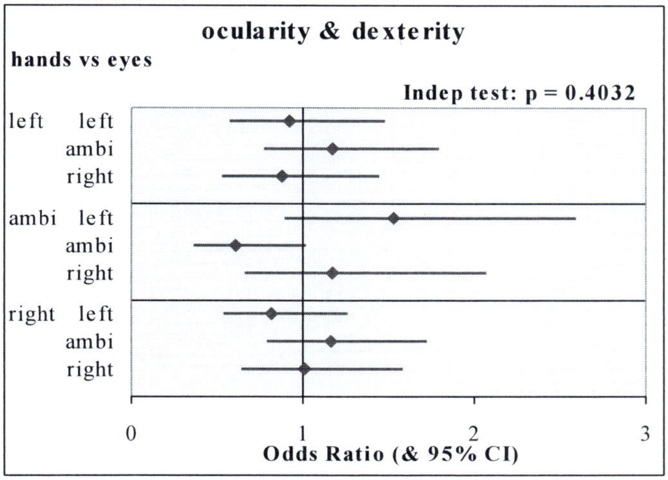

Figure 10.11: Ocularity-dexterity association among 413 subjects

Assumptions

Categorical data analysis does not require as many statistical assumptions as conventional statistical analysis based on continuous outcomes. The only assumption required (apart from the standard assumption of independent observations that underlies all basic statistical analysis) is that the counts in the contingency table be at least moderately large.

How large is 'moderately large'? It depends to some extent on the number of categories. However, the rule of thumb given in earlier chapters for the validity of the tests for population proportions is a useful yardstick, namely, that the expected count in each cell of the contingency table should be at least 5. Note that this condition applies to the expected counts rather than the observed counts.

Summary

In this chapter you have seen some methods for analysing categorical data. Categorical data are classified into specified groups, such as:

'male' or 'female' (gender);

'mild', 'moderate' or 'severe' (level of severity of a disease outcome);

'urban high-income', 'urban low-income', 'rural high-income' and 'rural low-income' (a demographic characteristic).

In the simplest case of categorical data there are only two categories. In this case the data are said to be binary.

Categorical data may be ordinal (if there is a natural ordering for the categories) or nominal (of there is no such ordering).

First, you saw how to compare a sample of categorical data with a given distribution. The method involves computing a chi-squared statistic based on comparing the observed frequencies with expected frequencies based on the given distribution. A p-value is then obtained for testing the null hypothesis that the population proportions are in agreement with this distribution. This p-value is the area in the upper tail of the chi-squared distribution. The chi-squared distribution is similar to the t distribution in one sense (it has a degrees of freedom parameter) but different in another sense (it is skewed, taking only non-negative values). The number of degrees of freedom is simply one less than the number of categories.

This method is called the chi-squared goodness-of-fit test. It is different from the r-squared goodness-of-fit described in Chapter 9.

Next, you saw how to test for the existence of an association between two categorical variables (a determinant and an outcome). This method again is based on a chi-squared statistic obtained by comparing the observed frequencies in a contingency table with expected frequencies. The test is called the chi-squared independence test. In this case the number of degrees of freedom is given by the product $(r-1)(c-1)$, where r is the number of categories for the determinant and c is the number of categories for the outcome.

If both the determinant and outcome variables are binary (in which case the data may be summarised as a 2×2 contingency table) the strength of the association may be measured by a quantity called the odds ratio. The odds associated with one of the two categories is simply the ratio $p/(1-p)$, where p is the proportion of data in the specified category, and the odds ratio is thus just the ratio of these odds for the two different levels of the determinant. Unlike probabilities whose values must lie between 0 and 1, odds can take any non-negative value. The odds ratio of two events is similar to their relative risk.

Finally, you saw how confidence intervals for odds ratios could be computed and graphed. These graphs may be extended to cover more general contingency tables (not just 2×2 contingency tables), and provide an informative summary of the association.

The only statistical assumption required for these methods to apply is that the sizes of the expected numbers of frequencies should not be too small. As a rule of thumb all expected counts should be no less than 5.

Homework problems

1. Library records in a certain tertiary institution have shown a long-term pattern that 70% of books are returned on time, 20% within one week of the due date, and the remainder after one week (or never). A random sample comprising 1000 loan records for the most recent year finds that 736 books were returned on time, 173 no more than a week late, and 91 more than a week late. Does this sample indicate a change of the pattern of returning loans?

2. A gynaecologist observed that 30 women out of 60 who gave birth at a particular hospital elected to have a complete spinal anesthetic during birth, and that 20 of their babies subsequently developed colic, compared with only 10 in the other group. Should the doctor conclude that there is an association?

3. On 19 May 1993 the *Sydney Morning Herald* reported that there were 46 sporting injuries on the previous Saturday which required the person being taken to hospital. Of 21 due to football injuries, 11 occurred to the head, neck or back. Of the non-football injuries there were 11 to the head, neck or back. Do these data show an association between playing football and suffering a head, neck or back injury? Answer the question by computing and plotting a 95% confidence interval for an appropriate odds ratio, and giving a p-value for an appropriate null hypothesis.

4. Consider again the study by Warton & Goodnow (1991) cited in Homework Problem 3 of Chapter 8. This study included 22 girls and 22 boys aged 14. For the selected tasks 'make own breakfast', 'pick up dirty clothes', and 'wash car', do these data indicate gender differences with respect to any of these tasks? Answer the question by showing an appropriate graph containing separate 95% confidence intervals of three appropriate odds ratios, and commenting on this graph.

5. In an article printed in *The Sydney Morning Herald* on 12 July 1993 the numbers of cash transactions of more than $10,000 were reported as follows.

	Significant transactions		Suspect transactions	
Cash dealer	1991/92	1992/93	1991/92	1992/93
Casino	3868	4678	68	42
TAB	1099	918	39	56
Bookmaker	602	396	0	0

   a. Considering only the Casino data, has the number of suspect transactions as a proportion of total cash transactions changed from the 1991/92 financial year to the 1992/93 financial year? (Assume that 'significant' cash transactions do not include 'suspect' cash transactions.)

   b. Consider the data for the 1992/93 financial year. Is there an association between the type of cash dealer and whether the transaction is suspect?

6. A report released by the New South Wales Department of Health claimed that "the campaign to teach parents not to let their babies sleep on their stomachs is believed to have halved the number of deaths from Sudden Infant Death Syndrome (SIDS) in the last two years" (*The Sydney Morning Herald*. 28 August 1993). In New South Wales in 1990, 185 babies died from SIDS, whilst in 1992, 79 died from the disease. Assess the Department's claim using the following methods:

   a. a chi-squared goodness-of-fit test;

   b. a graph showing a 95% confidence interval for the odds ratio, with the null hypothesis value indicated as a vertical dotted line.

   (For part b. assume that the populations at risk in 1990 and 1992 are the same, and that both populations are very large.)

7. To ascertain the level of comprehension of the greenhouse effect, Henderson-Sellers (1990) interviewed persons attending two national meetings on this issue. One of the questions they were asked involved the extent to which they believed that the ozone hole and the greenhouse effect were related. Of 120 attendees who were identified as having an 'accurate and good scientific understanding' of the issue, 8.5% said that they believed that the two phenomena were not at all related, 25.5% said that they were slightly related, 35.1% said somewhat related, 27.7% said very related, and 3.2% gave no response. For the other 201 respondents, the corresponding percentages were 2.5, 16.9, 32.8, 43.8, and 4. Is there evidence that the two groups differ in their responses?

8. Cotton, Antill & Cunningham (1990) sought to derive and determine the usefulness of an index representing mothers' attachment to the workforce. The proportions of families they targeted from each of three social risk areas were 35% from a 'low-risk' group, 27% from a 'medium-risk' group, and 38% from a 'high-risk' group. In their study of 166 families, it turned out that 68 came from the low-risk group, 48 from the medium-risk group, and the remainder from the high-risk group. Allowing for sampling error, did they achieve their targeted percentages?

9. For the study referred to in Homework Problem 8, the following table shows the outcomes in terms of the numbers of women in various employment classes.

Current work status	Work attachment *high*	*moderate*	*low*	total
Full-time employed	33	10	2	45
Part-time employed	11	36	3	50
Not employed	2	6	43	51
Looking for work	2	9	9	20

a. On the basis of this sample are the percentages of people falling into the three levels of work attachment the same?

b. Analyse the data to test if work attachment is independent of work status.

c. Display the data using an appropriate plot of odds ratios.

10. Pagano & Gauvreau (1993, page 133) cited data for 6,168 women who participated in the first U.S. National Health and Nutrition Examination Survey. The outcome variable of interest was whether or not the woman had developed breast cancer, and the determinant of interest was whether or not she had given birth to a child before age 25. Of 96 women who had developed breast cancer, 65 had given birth before age 25, while 4,475 of the women who had not developed breast cancer had given birth before age 25.
    a. Test the null hypothesis of no association.
    b. Graph the result as an odds ratio, with a 95% confidence interval.
    c. State your conclusion in words.

11. MacArthur (1931) gave data from a genetics experiment of tomato plants involving a dihybrid cross where the expected ratio of the four phenotypes listed is 9:3:3:1 (assuming no linkage). The numbers of offspring observed were as follows:

Phenotype	frequency
Tall cut-leaf	926
Tall, potato-leaf	288
Dwarf, cut-leaf	293
Dwarf, potato-leaf	104
Total	1611

Use a chi-squared goodness of fit test to compare the observed frequencies with the genetic theory.

APPENDICES

Appendix A: Tail Areas of Distributions

For convenience with computing p-values associated with the various tests of hypotheses described in Chapters 7-10, this Appendix contains the four tables of standardised distributions used.

Table A1 gives the area in a single tail, p, of the standardised normal distribution corresponding to the magnitude of a specified *z* score.

Table A2 gives the magnitude of a *t* statistic for a specified p-value, corresponding to the area in the two tails of the *t* distribution with the specified number of degrees of freedom.

Table A3 gives the value of a chi-squared statistic for a specified p-value, corresponding to the area in the upper tail of the chi-squared distribution with the specified number of degrees of freedom.

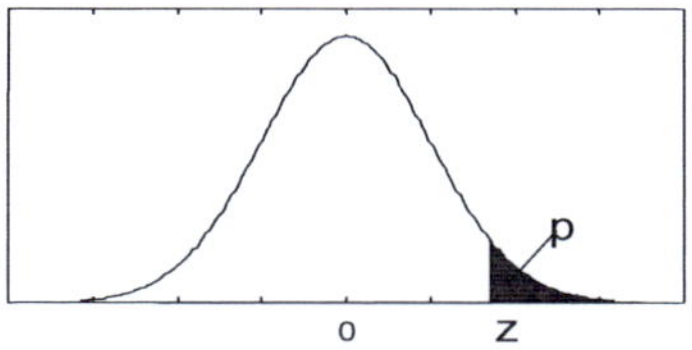

|z|	0.00	0.01	0.02	0.03	0.04	0.05	0.06	0.07	0.08	0.09
0.0	.5000	.4960	.4920	.4880	.4840	.4801	.4761	.4721	.4681	.4641
0.1	.4602	.4562	.4522	.4483	.4443	.4404	.4364	.4325	.4286	.4247
0.2	.4207	.4168	.4129	.4090	.4052	.4013	.3974	.3936	.3897	.3859
0.3	.3821	.3783	.3745	.3707	.3669	.3632	.3594	.3557	.3520	.3483
0.4	.3446	.3409	.3372	.3336	.3300	.3264	.3228	.3192	.3156	.3121
0.5	.3085	.3050	.3015	.2981	.2946	.2912	.2877	.2843	.2810	.2776
0.6	.2743	.2709	.2676	.2643	.2611	.2578	.2546	.2514	.2483	.2451
0.7	.2420	.2389	.2358	.2327	.2296	.2266	.2236	.2206	.2177	.2148
0.8	.2119	.2090	.2061	.2033	.2005	.1977	.1949	.1922	.1894	.1867
0.9	.1841	.1814	.1788	.1762	.1736	.1711	.1685	.1660	.1635	.1611
1.0	.1587	.1562	.1539	.1515	.1492	.1469	.1446	.1423	.1401	.1379
1.1	.1357	.1335	.1314	.1292	.1271	.1251	.1230	.1210	.1190	.1170
1.2	.1151	.1131	.1112	.1093	.1075	.1056	.1038	.1020	.1003	.0985
1.3	.0968	.0951	.0934	.0918	.0901	.0885	.0869	.0853	.0838	.0823
1.4	.0808	.0793	.0778	.0764	.0749	.0735	.0721	.0708	.0694	.0681
1.5	.0668	.0655	.0643	.0630	.0618	.0606	.0594	.0582	.0571	.0559
1.6	.0548	.0537	.0526	.0516	.0505	.0495	.0485	.0475	.0465	.0455
1.7	.0446	.0436	.0427	.0418	.0409	.0401	.0392	.0384	.0375	.0367
1.8	.0359	.0351	.0344	.0336	.0329	.0322	.0314	.0307	.0301	.0294
1.9	.0287	.0281	.0274	.0268	.0262	.0256	.0250	.0244	.0239	.0233
2.0	.0228	.0222	.0217	.0212	.0207	.0202	.0197	.0192	.0188	.0183
2.1	.0179	.0174	.0170	.0166	.0162	.0158	.0154	.0150	.0146	.0143
2.2	.0139	.0136	.0132	.0129	.0125	.0122	.0119	.0116	.0113	.0110
2.3	.0107	.0104	.0102	.0099	.0096	.0094	.0091	.0089	.0087	.0084
2.4	.0082	.0080	.0078	.0075	.0073	.0071	.0069	.0068	.0066	.0064
2.5	.0062	.0060	.0059	.0057	.0055	.0054	.0052	.0051	.0049	.0048
2.6	.0047	.0045	.0044	.0043	.0041	.0040	.0039	.0038	.0037	.0036
2.7	.0035	.0034	.0033	.0032	.0031	.0030	.0029	.0028	.0027	.0026
2.8	.0026	.0025	.0024	.0023	.0023	.0022	.0021	.0021	.0020	.0019
2.9	.0019	.0018	.0018	.0017	.0016	.0016	.0015	.0015	.0014	.0014
3.0	.00135	.00131	.00126	.00122	.00118	.00114	.00111	.00107	.00104	.00100
3.1	.00097	.00094	.00090	.00087	.00084	.00082	.00079	.00076	.00074	.00071
3.2	.00069	.00066	.00064	.00062	.00060	.00058	.00056	.00054	.00052	.00050
3.3	.00048	.00047	.00045	.00043	.00042	.00040	.00039	.00038	.00036	.00035
3.4	.00034	.00032	.00031	.00030	.00029	.00028	.00027	.00026	.00025	.00024
3.5	.00023	.00022	.00022	.00021	.00020	.00019	.00019	.00018	.00017	.00017
3.6	.00016	.00015	.00015	.00014	.00014	.00013	.00013	.00012	.00012	.00011
3.7	.00011	.00010	.00010	.00010	.00009	.00009	.00008	.00008	.00008	.00008
3.8	.00007	.00007	.00007	.00006	.00006	.00006	.00006	.00005	.00005	.00005
3.9	.00005	.00005	.00004	.00004	.00004	.00004	.00004	.00004	.00003	.00003
4.0	.000032									

Table A1: Single tail areas corresponding to z-values for the standardised normal curve

Two-tailed

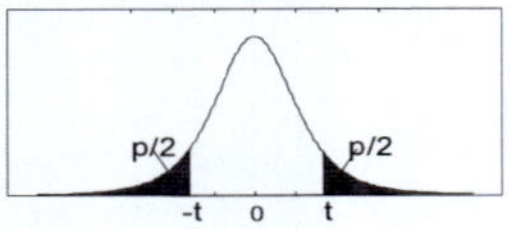

One-tailed

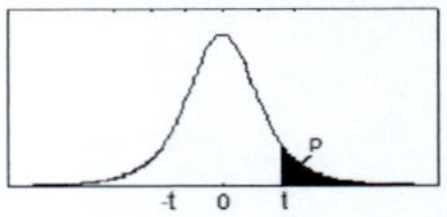

Two-tailed One-tailed ν	0.0005 0.00025	0.001 0.0005	0.002 0.001	0.005 0.0025	0.01 0.005	0.02 0.01	0.05 0.025	0.1 0.05	0.2 0.1	0.5 0.25
1	1273	636.6	318.3	127.3	63.66	31.82	12.71	6.314	3.078	1.000
2	44.70	31.60	22.33	14.09	9.925	6.965	4.303	2.920	1.886	0.816
3	16.33	12.92	10.22	7.453	5.841	4.541	3.182	2.353	1.638	0.765
4	10.31	8.610	7.173	5.598	4.604	3.747	2.776	2.132	1.533	0.741
5	7.976	6.869	5.893	4.773	4.032	3.365	2.571	2.015	1.476	0.727
6	6.788	5.959	5.208	4.317	3.707	3.143	2.447	1.943	1.440	0.718
7	6.082	5.408	4.785	4.029	3.499	2.998	2.365	1.895	1.415	0.711
8	5.617	5.041	4.501	3.833	3.355	2.896	2.306	1.860	1.397	0.706
9	5.291	4.781	4.297	3.690	3.250	2.821	2.262	1.833	1.383	0.703
10	5.049	4.587	4.144	3.581	3.169	2.764	2.228	1.812	1.372	0.700
11	4.863	4.437	4.025	3.497	3.106	2.718	2.201	1.796	1.363	0.697
12	4.716	4.318	3.930	3.428	3.055	2.681	2.179	1.782	1.356	0.696
13	4.597	4.221	3.852	3.372	3.012	2.650	2.160	1.771	1.350	0.694
14	4.499	4.140	3.787	3.326	2.977	2.624	2.145	1.761	1.345	0.692
15	4.417	4.073	3.733	3.286	2.947	2.602	2.131	1.753	1.341	0.691
16	4.346	4.015	3.686	3.252	2.921	2.583	2.120	1.746	1.337	0.690
17	4.286	3.965	3.646	3.222	2.898	2.567	2.110	1.740	1.333	0.689
18	4.233	3.922	3.610	3.197	2.878	2.552	2.101	1.734	1.330	0.688
19	4.187	3.883	3.579	3.174	2.861	2.539	2.093	1.729	1.328	0.688
20	4.146	3.850	3.552	3.153	2.845	2.528	2.086	1.725	1.325	0.687
21	4.110	3.819	3.527	3.135	2.831	2.518	2.080	1.721	1.323	0.686
22	4.077	3.792	3.505	3.119	2.819	2.508	2.074	1.717	1.321	0.686
23	4.047	3.768	3.485	3.104	2.807	2.500	2.069	1.714	1.319	0.685
24	4.021	3.745	3.467	3.091	2.797	2.492	2.064	1.711	1.318	0.685
25	3.996	3.725	3.450	3.078	2.787	2.485	2.060	1.708	1.316	0.684
26	3.974	3.707	3.435	3.067	2.779	2.479	2.056	1.706	1.315	0.684
27	3.954	3.690	3.421	3.057	2.771	2.473	2.052	1.703	1.314	0.684
28	3.935	3.674	3.408	3.047	2.763	2.467	2.048	1.701	1.313	0.683
29	3.918	3.659	3.396	3.038	2.756	2.462	2.045	1.699	1.311	0.683
30	3.902	3.646	3.385	3.030	2.750	2.457	2.042	1.697	1.310	0.683
35	3.836	3.591	3.340	2.996	2.724	2.438	2.030	1.690	1.306	0.682
40	3.788	3.551	3.307	2.971	2.704	2.423	2.021	1.684	1.303	0.681
45	3.752	3.520	3.281	2.952	2.690	2.412	2.014	1.679	1.301	0.680
50	3.723	3.496	3.261	2.937	2.678	2.403	2.009	1.676	1.299	0.679
60	3.681	3.460	3.232	2.915	2.660	2.390	2.000	1.671	1.296	0.679
70	3.651	3.435	3.211	2.899	2.648	2.381	1.994	1.667	1.294	0.678
80	3.629	3.416	3.195	2.887	2.639	2.374	1.990	1.664	1.292	0.678
90	3.612	3.402	3.183	2.879	2.632	2.368	1.987	1.662	1.291	0.677
100	3.598	3.390	3.174	2.871	2.626	2.364	1.984	1.660	1.290	0.676
∞	3.481	3.291	3.090	2.807	2.576	2.326	1.960	1.645	1.282	0.674

Table A2: Values of $|t|$ corresponding to p-values for Student's t distribution

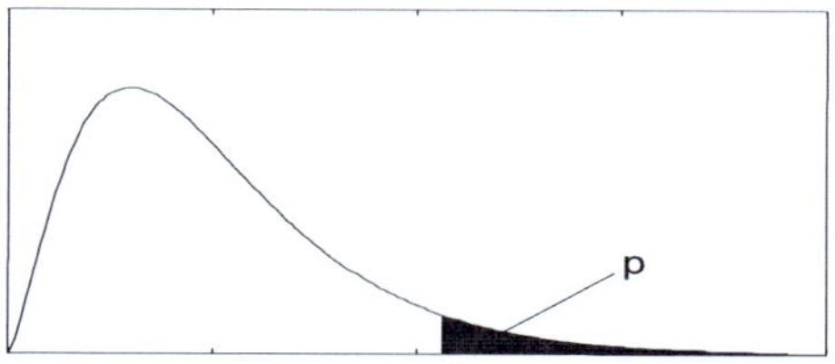

ν	.0005	0.001	0.002	0.005	0.01	0.02	0.05	0.1	0.2	0.5
1	12.12	10.83	9.550	7.879	6.635	5.412	3.842	2.706	1.642	0.455
2	15.20	13.82	12.43	10.60	9.210	7.824	5.992	4.605	3.219	1.386
3	17.73	16.27	14.80	12.84	11.34	9.837	7.815	6.251	4.642	2.366
4	20.00	18.47	16.92	14.86	13.28	11.67	9.488	7.779	5.989	3.357
5	22.11	20.52	18.91	16.75	15.09	13.39	11.07	9.236	7.289	4.352
6	24.10	22.46	20.79	18.55	16.81	15.03	12.59	10.64	8.558	5.348
7	26.02	24.32	22.60	20.28	18.48	16.62	14.07	12.02	9.803	6.346
8	27.87	26.12	24.35	21.95	20.09	18.17	15.51	13.36	11.03	7.344
9	29.67	27.88	26.06	23.59	21.67	19.68	16.92	14.68	12.24	8.343
10	31.42	29.59	27.72	25.19	23.21	21.16	18.31	15.99	13.44	9.342
11	33.14	31.26	29.35	26.76	24.72	22.62	19.68	17.28	14.63	10.34
12	34.82	32.91	30.96	28.30	26.22	24.05	21.03	18.55	15.81	11.34
13	36.48	34.53	32.54	29.82	27.69	25.47	22.36	19.81	16.98	12.34
14	38.11	36.12	34.09	31.32	29.14	26.87	23.68	21.06	18.15	13.34
15	39.72	37.70	35.63	32.80	30.58	28.26	25.00	22.31	19.31	14.34
16	41.31	39.25	37.15	34.27	32.00	29.63	26.30	23.54	20.47	15.34
17	42.88	40.79	38.65	35.72	33.41	31.00	27.59	24.77	21.61	16.34
18	44.43	42.31	40.14	37.16	34.81	32.35	28.87	25.99	22.76	17.34
19	45.97	43.82	41.61	38.58	36.19	33.69	30.14	27.20	23.90	18.34
20	47.50	45.31	43.07	40.00	37.57	35.02	31.41	28.41	25.04	19.34
21	49.01	46.80	44.52	41.40	38.93	36.34	32.67	29.62	26.17	20.34
22	50.51	48.27	45.96	42.80	40.29	37.66	33.92	30.81	27.30	21.34
23	52.00	49.73	47.39	44.18	41.64	39.97	35.17	32.01	28.43	22.34
24	53.48	51.18	48.81	45.56	42.98	40.27	36.42	33.20	29.55	23.34
25	54.95	52.62	50.22	46.93	44.31	41.57	37.65	34.38	30.68	24.34
26	56.41	54.05	51.63	48.29	45.64	42.86	38.89	35.56	31.79	25.34
27	57.86	55.48	53.02	49.64	46.96	44.14	40.11	36.74	32.91	26.34
28	59.30	56.89	54.41	50.99	48.28	45.42	41.34	37.92	34.03	27.34
29	60.73	58.30	55.79	52.34	49.59	46.69	42.56	39.09	35.14	28.34
30	62.16	59.70	57.17	53.67	50.89	47.96	43.77	40.26	36.25	29.34
35	69.20	66.62	63.95	60.27	57.34	54.24	49.80	46.06	41.78	34.34
40	76.09	73.40	70.62	66.77	63.69	60.44	55.76	51.81	47.27	39.34
45	82.88	80.08	77.18	73.17	69.96	66.56	61.66	57.51	52.73	44.34
50	89.56	86.66	83.66	79.49	76.15	72.61	67.50	63.17	58.16	49.33
60	102.7	99.61	96.40	91.95	88.38	84.58	79.08	74.40	68.97	59.33
70	115.6	112.3	108.9	104.2	100.4	96.39	90.53	85.53	79.71	69.33
80	128.3	124.8	121.3	116.3	112.3	108.1	101.8	96.58	90.41	79.33
90	140.8	137.2	133.5	128.3	124.1	119.6	113.1	107.6	101.0	89.33
100	153.2	149.4	145.6	140.2	135.8	131.1	124.3	118.5	111.7	99.33

Table A3: Values of χ^2 corresponding to p-values for the chi-squared distribution

Appendix B: 1988 Student Survey Data

This appendix contains a list of some data collected from a survey of students attending lectures in Stat170 *Introductory Statistics* at Macquarie University on 25 May 1988. A questionnaire was handed out in each class and collected after a few minutes. 426 students returned responses that could be evaluated, a response rate of over 80%. The questionnaire form is shown on the next page, with the complete list of responses on the following pages. Of the 426 students who responded, 270 said that they were born in New South Wales. The data set was reduced to one with 183 cases by excluding 90% of the students born in New South Wales, but including all the others. This set of data is the one frequently referenced in the text, and corresponds to the first 183 cases listed.

Besides providing an interesting source of data that could be analysed by students enrolled in Stat170 in later years, the object of the survey was to obtain some feedback about the relative workload and the quality of teaching in the unit. A linear analogue scale of assessment (LASA) scale was used for each of four response variables. These responses were coded by dividing the scale into ten equal divisions and coding each response as an integer from 0 to 9, this being the number of divisions to the left of where the student marked their response. (Responses marked outside the range were truncated.)

You may like to consider handing out a similar questionnaire in your current classes and compare the responses to those obtained in 1988. For example, you may find a change in smoking habits or favourite drink. You would also need to change the expected grade to HD, D, Cr, P or F as the grades at Macquarie University changed in 2000.

The data were not subjected to any serious quality control checks, and thus contain errors.

1988 MACQUARIE UNIVERSITY STUDENT CENSUS

I am ☐ Male ☐ Female I was born in ☐☐ (year) ☐☐ (month)

My height is ☐☐☐ (cm) My weight is ☐☐☐ (kg, approx)

My marital status is: **1** never married **2** married **3** other (circle one)

I was born in ____________________ (Country, or State if Australia)

My favourite drink on social occasions is ______________________

The discipline I plan to major in is: ______________________
(write "unsure" if not known yet)

My expected grade in STAT170 is ☐ A ☐ B ☐ C ☐ CQ ☐ F

(Please put crosses in appropriate places on each scale)

Compared to other courses I'm taking, or have taken, at Macquarie

the workload on STAT170 is **light** |————| **heavy**

the quality of teaching is **poor** |————| **excellent**

my learning progress is **slow** |————| **rapid**

the course is **boring** |————| **exciting**

I have a car ☐ Yes ☐ No I have (or share) a home computer ☐ Yes ☐ No

I travel _________ (km) to get to Macquarie.

I attend a religious service on _________ days per year (approx)

I smoke _______ cigarettes/day (approx..)

I watch TV for ________ hours/week (approx.)

ID	sex	yr	mon	ht	wt	ms	origin	drink	discip	gr	w	q	l	i	v	c	dist	pray	sm	TV
1	f	70	7	165	60	1	ACT	Orange J	Psych	C	6	3	3	3	n	y	15	0	0	17
2	f	69	8	175	70	1	ACT	Tequila	Educ	B	6	1	4	4	n	n	0	5	15	0
3	m	70	7	185	66	3	ACT	Water	Mass	B	6	7	7	7	n	y	90	104	0	15
4	m	70	5	187	91	1	NSW	Water		A	9	6	6	4	y	n	32	356	0	0
5	m	70	5	180	70	1	NSW	Coke	Econ	C	8	3	1	0	y	n	25	0	0	20
6	m	69	5	179	76	1	NSW	Orgasm	Econ	C	4	2	3	0	n	n	25	1	0	20
7	m	68	9	172	66	1	NSW	Scotch	Econ	B	7	4	7	4	n	n	15	0	0	3
8	m	70	2	189	82	*	NSW	Russian F	Econ	A	7	2	9	0	y	y	15	0	0	7
9	f	68	12	178	65	1	NSW	Cider	Unsure	B	4	7	6	5	y	y	50	52	0	20
10	f	70	4	160	55	1	NSW	S.Comfort	Mass	CQ	9	5	0	0	n	n	0	0	0	0
11	f	56	8	165	75	2	NSW	Mineral W	Engl	C	8	7	7	5	y	y	10	0	0	2
12	f	69	11	155	45	1	NSW	Kalua	BHS	C	8	3	4	1	n	n	10	5	0	2
13	m	68	9	185	70	1	NSW	Beer	Comp	C	4	4	7	2	n	n	25	23	0	20
14	f	68	10	167	67	1	NSW	Kalua	Econ	C	9	0	3	3	y	n	7	55	0	13
15	f	64	10	182	67	1	NSW	Brandy	Unsure	C	9	4	2	4	y	n	10	4	0	4
16	f	70	1	166	50	1	NSW	Lemonade	Soc	B	3	7	8	4	n	n	0	55	0	6
17	m	68	9	182	96	1	NSW	Bourbon	Accg	B	4	4	4	2	n	y	15	52	0	14
18	m	69	10	180	71	1	NSW	Beer	Econ	C	5	8	6	5	y	n	7	1	3	10
19	m	69	6	196	77	1	NSW	Vodka	Mass	C	7	4	3	1	y	y	8	30	0	*
20	f	70	2	173	*	1	NSW	Wine	Econ	C	5	3	5	1	n	y	2	55	0	15
21	m	70	7	172	70	1	NSW	Beer	Econ	C	9	2	3	0	n	n	10	0	0	15
22	f	68	6	168	63	1	NSW	Orgasm	Econ	B	7	9	8	6	y	n	23	0	0	7
23	m	69	8	178	65	1	NSW	Beer	Law	B	5	5	5	1	y		10	0	30	30
24	m	47	8	171	77	2	NSW	Orange J	Anthr	A	4	9	9	9	y	y	3	52	0	5
25	f	49	4	165	60	2	NSW	Mineral W	Psych	C	7	7	1	5	y	y	7	12	0	1
26	f	69	10	170	57	1	NSW	Snowball	Accg	C	4	3	3	3	n	n	20	0	0	0
27	f	70	2	175	68	1	NSW	Water	Econ	C	6	4	4	3	y	n	0	2	0	3
28	f	69	8	171	56	1	NSW	Coke	Psych	C	4	5	5	1	y	y	25	3	0	10
29	f	69	2	170	65	1	NSW	Vodka	Unsure	B	8	6	7	2	n	n	15	40	5	3
30	f	69	6	175	55	1	NSW	Avocado	Unsure	B	6	6	3	2	n	n	20	0	0	2
31	f	57	5	153	54	2	Qld	Bacardi	Econ	B	7	7	6	7	y	n	25	1	0	3
32	m	70	1	180	77	1	Qld	S.Comfort	Accg	B	4	2	1	6	y	y	15	0	0	20
33	f	69	4	170	57	1	Qld	Squash	Biol	B	4	3	7	2	n	y	1	60	0	2
34	f	50	8	162	57	2	Qld	Wine	Psych	A	2	6	6	1	n	y	22	0	0	8
35	f	40	2	157	49	2	Qld	Wine	Ling	CQ	8	6	2	4	y	n	11	0	25	2
36	m	69	11	180	74	1	Qld	Beer	Econ	B	7	6	8	3	y	y	10	3	0	6
37	f	70	4	175	55	1	Qld	Scotch	Mass	B	6	8	4	0	n	n	25	2	0	6
38	m	69	8	175	70	1	Qld	Cointreau	Mass	C	7	7	6	4	n	n	0	1	0	3
39	f	62	2	172	69	1	S.Aust	Champagne	ENGL	C	8	4	3	2	y	n	15	1	25	1
40	m	70	6	190	77	1	S.Aust	Guinness	Mass	A	1	4	6	0	n	n	0	0	0	4
41	f	70	9	173	56	1	S.Aust	Gin	Educ	C	6	4	4	4	n	n	70	60	0	10

ID	sex	yr	mon	ht	wt	ms	origin	drink	discip	gr	w	q	l	i	v	c	dist	pray	sm	TV
42	f	64	6	164	60	1	S.Aust	Coke	Sci	C	9	0	3	0	y	n	18	0	0	10
43	f	69	9	175	56	1	S.Aust	Bacardi	Educ	CQ	7	2	2	0	y	y	10	0	0	8
44	f	69	7	168	70	1	Vic	Scotch	Econ		6	4	4	1	y	y	1	10	20	21
45	m	70	4	176	60	1	Vic	Wine	Stats	B	8	6	6	6	y	n	4	1	0	10
46	f	68	1	175	56	1	Vic	Vodka	Mass	B	5	6	5	4	y	n	15	60	0	10
47	f	70	3	170	56	*	Vic	Soft Dr	Unsure	C	8	3	5	1	n	y	10	3	15	8
48	f	70	7	178	55	1	Vic	Fluffy D	Psych	C	4	4	1	0	n	y	0	0	0	1
49	f	70	6	170	60	1	W.Aust	S.Comfort	Educ	C	8	7	4	2	y	n	15	0	0	12
50	f	48	7	125	54	2	W.Aust	Champagn	Psych	C	5	5	5	5	y	n	10	20	0	3
51	f	61	12	180	62	2	W.Aust	Wine	Psych	B	7	4	4	3	y	n	13	20	0	10
52	m	70	4	172	65	1	Brunei	Vodka	Stat/C	B	5	3	5	2	n	n	20	0	0	25
53	m	66	3	174	73	1	Indon.	Coke	Comp	B	6	5	5	6	y	y	11	2	0	21
54	m	65	9	177	71	1	Indon.	Beer	Accg	B	4	4	4	3	y	n	18	30	5	15
55	f	69	9	160	50	1	Indon.	Lemonade		B	4	1	5	4	n	n	1	52	0	1
56	f	68	2	155	45	1	Indon.	Orange J	Unsure	B	9	4	1	9	n	n	0	52	0	5
57	m	68	8	174	68	1	Indon.	Coke	Comp	A	4	5	6	5	y	y	2	0	0	14
58	m	67	5	167	52	1	Indon.	Whisky	Stat/C	A	5	3	5	5	n	y	22	0	0	10
59	m	66	12	174	79	1	Indon.	Orange J	Stat/C	B	7	3	9	6	n	n	25	104	0	12
60	m	62	4	165	68	1	Kamp.	Soft Dr	Unsure	B	4	3	4	4	y	n	27	0	0	5
61	f	67	12	163	58	1	Laos	Soft Dr	Unsure	C	5	5	5	5	n	n	0	0	0	10
62	f	69	6	160	48	1	Laos	Apple J	Unsure	B	6	4	4	4	n	n	0	40	0	10
63	f	70	1	160	55	1	Malay.	Red Wine	Econ	B	8	3	5	5	n	y	50	50	0	7
64	f	69	10	160	52	1	Malay.	Orange J	Accg	B	6	8	5	5	n	n	30	54	0	10
65	m	68	10	175	66	1	Malay.	Juice	Accg	B	6	2	4	3	y	n	20	0	0	3
66	f	70	3	160	60	1	Malay.	Lemonade	Unsure	C	6	4	4	4	n	y	0	0	0	6
67	m	70	5	175	65	1	Malay.	Bourbon	Econ	A	7	7	8	4	y	y	3	50	0	35
68	m	69	4	183	80	1	Malay.	Samboka	Comp	C	9	4	6	6	y	y	30	0	0	8
69	m	65	11	174	60	1	Malay.	Orange J	Comp	B	8	2	5	6	n	n	12	0	0	10
70	f	67	8	156	43	*	Malay.	Ginger B	Econ	C	7	6	5	3	n	n	10	0	0	10
71	f	68	11	*	47	3	Malay.		Unsure	C	5	5	3	6	n	n	*	1	0	33
72	m	68	7	178	61	1	Malay.	Champagne	Accg	B	8	6	6	4	n	y	20	60	0	16
73	m	68	7	170	55	1	Malay.	Apple J	Accg	A	2	8	8	6	n	n	0	100	0	2
74	m	69	11	170	61	1	Malay.	Cider	Econ		7	5	2	5	y	n	15	2	0	9
75	f	70	3	179	60	*	Malay.	S.Comfort	Accg	B	6	4	5	0	n	n	0	20	0	2
76	m	69	7	176	70	1	Malay.	Coke	Law	B	7	6	6	6	n	n	10	30	0	7
77	f	68	7	155	51	3	Malay.	Coke	Unsure	C	7	4	5	4	n	n	2	1	0	3
78	m	66	7	170	68	1	Philip	Beer	Mass	A	4	4	6	4	n	n	12	40	0	6
79	m	70	1	187	65	1	Philip	Soft Dr	Mass	B	4	4	5	4	n	n	38	40	0	7
80	f	68	9	143	47	1	Philip	Ginger B	Stats	C	5	4	4	4	n	n	4	40	0	*
81	f	63	6	157	48	1	Singap	Fruit J	Accg	B	4	4	4	3	n	n	17	48	0	10
82	m	67	10	170	58	1	Viet.	Coffee	Econ	B	5	8	5	7	n	n	30	0	0	10

ID	sex	yr	mon	ht	wt	ms	origin	drink	discip	gr	w	q	l	i	v	c	dist	pray	sm	TV
83	f	65	12	152	40	*	Viet.		Comp	C	8	5	2	5	n	n	30	0	0	6
84	f	60	1	156	39	1	Viet.	None	Unsure	C	9	0	0	4	n	n	30	0	0	10
85	m	69	2	175	75	1	Viet.	Red Wine	Comp	C	6	6	4	3	y	n	49	0	10	10
86	f	67	2	162	57	1	Viet.	Orange J	Unsure	B	6	6	4	6	n	n	0	0	0	6
87	f	69	3	160	59	1	Viet.	Champagne	Psych	CQ	4	3	4	0	n	n	20	0	0	5
88	f	66	7	154	45	1	Viet.		Comp		5	4	4	3	n	n	0	0	0	*
89	m	61	5	180	79	*	China	Beer	Unsure	B	4	7	7	6	y	n	20	0	0	5
90	f	62	12	162	46	1	Hong K	Orange J	Comp	B	4	6	6	8	y	n	20	0	0	5
91	m	65	6	180	72	1	Hong K	Coke	Comp	B	4	9	7	7	n	y	30	52	0	2
92	m	67	7	178	75	1	Hong K	Beer	Unsure	B	9	6	5	7	y	y	40	0	0	10
93	m	68	11	175	65	1	Hong K	Daiquiri	Comp	B	8	3	9	1	y	y	30	30	0	8
94	m	69	5	175	65	1	Hong K		Accg	A	8	4	5	5	n	n	10	55	0	10
95	f	67	2	170	55	1	Hong K	Coke		CQ	7	5	5	5	y	n	40	20	0	2
96	f	69	8	165	50	1	Hong K	Lemon Sq	Comp	B	7	6	6	5	n	n	30	80	0	6
97	f	69	9	162	42	1	Hong K	Soft Dr	Econ	C	5	4	2	5	y	n	6	1	0	10
98	f	69	2	166	55	1	Hong K		Unsure	C	4	4	3	1	n	n	*	52	0	*
99	m	69	9	176	55	1	Hong K	Bacardi		B	8	4	5	1	n	y	10	1	0	0
100	m	65	12	185	62	1	Hong K	Coke	Comp	C	4	4	3	4	y	n	25	0	0	2
101	m	68	4	170	57	1	Hong K	Gin	Comp	B	8	2	9	1	y	n	7	0	0	18
102	m	69	10	176	78	1	Hong K	Zombie	Comp	B	7	3	9	2	y	y	12	0	0	30
103	m	67	8	175	61	1	Hong K	Coke	Unsure	A	3	3	6	2	y	n	0	3	0	5
104	f	61	5	150	50	1	Hong K	Orange J	Accg	C	9	2	2	1	y	n	8	0	0	6
105	m	70	1	180	73	1	Hong K	Wine	Unsure	A	4	4	8	1	n	y	2	60	0	15
106	f	65	12	160	62	1	Hong K	Water	Mass	B	9	2	2	2	n	n	50	78	0	16
107	f	69	6	159	45	1	Hong K	Orange J	Unsure	B	5	3	4	4	y	y	23	52	0	8
108	f	69	8	157	47	1	India	Orange J		A	8	5	9	1	n	n	15	1	0	7
109	f	65	8	160	48	1	India	Solo	Comp	B	4	2	7	6	y	y	15	0	10	5
110	f	70	8	157	49	1	India		Unsure	CQ	5	2	2	5	n	n	30	0	0	8
111	m	69	9	182	75	1	India	Bourbon	Comp/S	A	7	7	9	9	n	n	15	65	0	12
112	f	66	6	162	49	2	India	Benedict	Psych	C	9	7	6	3	y	n	34	0	20	3
113	f	69	10	165	*	1	Iran	Wine	Econ	B	9	4	6	0	n	n	15	0	0	1
114	f	68	6	165	50	1	Iran	Orange J	Maths	C	5	5	5	2	n	n	8	2	0	7
115	m	65	8	185	80	1	Iran	Beer	Law	B	0	9	0	0	n	n	20	1	15	6
116	f	61	8	160	51	1	Iraq	Lemon Sq	Maths	C	8	6	5	4	y	y	25	30	0	5
117	f	69	6	165	60	1	Israel	Anything	Mass	CQ	9	0	0	0	n	n	10	0	0	8
118	f	64	12	164	50	1	Japan		Psych	B	9	6	6	5	y	n	15	1	0	10
119	m	69	1	179	78	1	Japan		Unsure	B	9	2	4	2	y	n	25	0	0	10
120	m	68	5	182	65	1	Korea	Beer	Unsure	B	3	5	6	4	n	n	15	52	0	4
121	f	70	7	157	47	1	Korea	Tia Maria	Psych	B	6	1	1	0	n	n	90	0	0	14
122	f	69	6	160	53	1	Korea	Perrier	Econ	B	5	5	4	1	n	n	20	0	0	30
123	f	63	11	152	55	3	Leban.	Kalua	Mass	C	8	2	2	1	y	n	20	4	0	10

ID	sex	yr	mon	ht	wt	ms	origin	drink	discip	gr	w	q	l	i	v	c	dist	pray	sm	TV
124	f	69	12	165	60	1	Leban.	Tia Maria	Mass	C	8	8	1	4	y	n	60	50	0	8
125	m	65	2	171	67	1	Macao	Scotch	Maths	C	8	1	1	4	n	y	5	0	10	15
126	f	69	9	175	54	3	PNG	Coke	Psych	C	6	3	4	4	y	n	40	30	1	10
127	f	68	1	160	42	1	PNG	Lemonade	Biol	C	4	5	5	3	y	y	17	*	0	10
128	f	70	5	157	*	1	PNG	Orange J	Psych	C	6	3	3	3	y	y	15	2	0	25
129	f	69	12	170	60	1	PNG	S.Comfort	Unsure	C	9	6	1	3	y	y	15	52	0	42
130	f	69	4	155	60	1	PNG	Coke	Unsure	F	9	0	0	0	y	n	20	0	0	10
131	m	62	3	160	55	*	Sri L	Beer	Comp	B	4	7	6	7	n	n	8	48	0	7
132	m	70	2	175	65	1	Sri L	Beer		C	7	6	7	3	y	n	10	0	0	10
133	f	52	4	163	54	2	Czech	Wine	Psych	A	7	4	7	7	y	n	25	52	0	4
134	f	56	4	166	54	3	Denmar	Wine	Educ	A	8	6	6	6	y	n	15	0	0	0
135	m	66	9	185	78	1	Engl.	Beer	GEO	C	5	5	5	5	y	n	40	2	2	10
136	m	69	11	180	80	3	Engl.	Bourbon	Econ	B	4	5	5	6	n	y	10	1	0	14
137	m	70	1	173	65	1	Engl.	Beer	Econ	C	4	4	4	2	n	n	3	0	0	25
138	f	66	1	175	50	1	Engl.	Solo	Econ	C	6	2	4	2	n	n	7	0	15	8
139	m	52	11	172	60	3	Engl.	Wine	BHS	C	5	5	3	4	y	y	14	0	0	8
140	f	45	2	171	65	3	Engl.	Soft Dr	Psych	C	9	8	2	4	y	n	60	0	0	6
141	f	53	3	160	60	2	Engl.	Rum	Psych		9	5	5	4	y	n	30	1	0	2
142	m	69	1	180	80	3	Greece	Bacardi	Comp	C	9	5	5	5	y	y	10	0	0	7
143	m	26	10	174	73	2	Holl.	Scotch	Biol	B	4	9	3	9	y	n	9	60	0	6
144	f	70	4	165	55	1	Hung.	Coke	Comp	A	6	4	9	5	n	n	100	0	0	5
145	f	65	11	160	54	1	Italy	Mineral W	Unsure	C	8	4	1	5	y	n	40	0	3	1
146	f	68	5	167	53	*	Malta	Coffee	Biol	B	7	3	5	5	n	y	25	0	0	5
147	f	67	9	162	58	1	Poland	Wine	Mass	B	4	6	5	2	y	n	0	2	0	8
148	m	69	10	176	111	2	Russia	Tia Maria	Unsure	A	8	2	3	3	y	n	2	0	0	15
149	m	69	3	190	69	1	Scot.	Coopers	Econ	C	9	4	2	0	n	n	7	0	0	10
150	m	65	3	179	75	1	Scot.	Beer	Mass	B	1	6	9	1	n	n	7	0	0	8
151	f	68	9	171	55	3	Scot.	Spirits	Mass	B	4	7	3	1	n	n	0	2	0	8
152	m	70	7	194	105	*	Serbia	Coke	Stats	CQ	9	9	0	9	n	n	40	1	0	72
153	m	69	7	179	73	1	Sweden	Beer	Econ	B	6	3	5	2	n	n	15	0	0	9
154	m	69	7	180	75	1	Sweden	Beer	Unsure	C	9	2	4	1	n	n	16	0	0	7
155	m	69	6	183	60	1	UK	Water	Unsure	F	9	4	7	0	y	y	14	1	0	1
156	f	69	3	156	53	1	UK	Tia Maria	Engl	C	4	3	2	1	n	y	50	0	0	16
157	m	63	9	180	82	1	UK	Fruit J	GEO	B	4	7	4	6	y	y	8	48	0	10
158	m	69	9	180	70	1	Yugos.	Beer	Geog	C	8	1	6	0	y	y	9	0	0	10
159	f	60	8	178	68	1	Canada	Wine	Econ	C	7	4	4	2	n	n	20	3	15	9
160	f	66	11	165	50	1	Canada	Pimms	Psych	B	4	6	4	3	n	y	20	52	0	6
161	f	69	*	170	53	1	Chile	S.Comfort	Accg	C	7	5	6	1	n	n	25	5	0	25
162	m	63	11	173	70	1	Fiji	Scotch	Accg	C	9	4	4	1	y	n	22	10	0	3
163	m	68	9	175	68	1	Fiji	Soft Dr	Econ	B	5	9	8	5	n	n	0	15	0	0
164	f	69	10	155	48	1	Fiji	Gin	Engl	C	7	3	3	3	n	n	8	1	0	5

ID	sex	yr	mon	ht	wt	ms	origin	drink	discip	gr	w	q	l	i	v	c	dist	pray	sm	TV
165	f	67	5	172	55	2	Fiji	Tia Maria	Econ	C	5	4	5	6	n	n	50	365	0	5
166	m	62	10	165	50	2	Fiji	Beer	Unsure	B	7	2	1	6	n	n	0	1	0	2
167	m	67	3	170	60	1	Maurit	Soft Dr	Comp	B	4	4	4	2	n	n	10	0	0	10
168	f	43	5	162	54	2	Maurit	Wine	Psych	C	5	6	4	4	y	y	10	2	0	7
169	f	67	5	175	55	1	NZ	Kalua	Educ	C	7	3	3	3	n	n	1	2	0	15
170	f	63	7	156	58	1	NZ	S.Comfort	Psych	CQ	8	3	5	2	y	n	10	56	0	14
171	m	70	7	180	70	1	NZ	Vodka	Econ	B	3	5	5	3	n	y	45	0	0	15
172	m	70	1	165	68	1	S.Afr.		Unsure	B	4	7	7	9	n	n	30	0	0	10
173	m	69	12	168	82	1	S.Afr.	Kalua	Econ	A	2	5	8	9	y	y	30	22	0	24
174	f	68	3	160	63	1	S.Afr.			B	8	9	5	8	y	n	20	0	0	7
175	m	68	8	186	66	1	S.Afr.	Malibu	Mass	A	8	7	6	1	y	n	30	20	0	30
176	f	70	3	170	55	1	S.Afr.	Kalua	Accg	B	2	6	4	5	n	n	0	0	0	5
177	m	69	3	184	78	1	USA	Coopers	Econ	C	7	7	5	4	y	n	4	0	5	3
178	m	70	6	174	65	1	USA	Gin	Law	B	7	6	8	3	y	y	4	3	0	15
179	f	66	6	149	56	1	USA	Rum	Accg	B	8	4	6	3	n	n	6	10	0	14
180	m	66	12	166	55	1	USA	Beer	Comp	B	6	9	4	6	y	y	15	0	0	7
181	m	61	4	180	75	2	USA	Beer	Comp	B	6	4	3	5	y	y	30	5	0	20
182	f	69	3	163	48	1	USA	Wine	Mass	CQ	8	2	1	3	y	n	20	0	0	4
183	m	57	1	188	72	3	Vanuat	Iced W	Econ	A	0	4	9	0	y	y	35	60	0	2
184	f	68	11	157	46	1	NSW	Vodka	Unsure	B	6	5	4	5	n	n	25	0	0	10
185	f	70	4	175	55	1	NSW	Gin	ECFS	B	6	4	8	1	y	n	20	55	0	12
186	f	70	4	175	62	1	NSW	Bacardi	Accg	B	5	4	6	2	y	y	30	0	0	10
187	m	69	11	178	62	1	NSW	Beer	Unsure	C	4	3	4	3	n	y	20	0	0	10
188	f	70	6	172	55	*	NSW	Champagne	Unsure	B	6	6	6	1	n	n	30	10	0	20
189	f	70	1	165	53	1	NSW	Vodka	Mass	C	7	5	3	1	y	y	25	0	0	6
190	m	70	4	178	80	1	NSW	Vodka	Accg	C	8	3	9	0	y	y	23	0	0	20
191	f	69	6	165	57	1	NSW	Tia Maria	Accg	B	4	3	5	2	n	n	3	58	0	20
192	f	69	10	160	60	1	NSW	Champagne	Accg	B	4	4	4	4	n	n	4	52	0	10
193	m	70	7	182	75	1	NSW	Bourbon	Econ	A	1	4	7	5	y	n	27	52	0	12
194	f	69	7	177	69	1	NSW	Orange J	Accg	B	4	2	7	0	n	n	30	15	0	8
195	f	69	10	163	64	1	NSW	Mineral W	Unsure	C	7	3	1	4	n	n	30	15	0	8
196	f	70	3	163	60	3	NSW	Kalua	Unsure	B	7	4	5	2	y	y	12	0	0	20
197	m	60	9	178	79	2	NSW	Red Wine	Accg	A	4	5	3	2	y	y	30	0	0	15
198	m	69	10	180	70	1	NSW	Milk	Econ	A	4	5	6	0	n	n	30	0	0	0
199	f	70	2	174	60	1	NSW	Lemonade	Accg	A	6	4	6	5	y	n	30	4	0	0
200	m	69	10	173	73	1	NSW	Cooler	Accg	C	7	4	4	3	y	y	5	0	0	10
201	f	61	4	170	65	2	NSW	Bourbon	Anthr	C	9	0	3	3	y	y	80	0	20	10
202	m	70	5	175	70	1	NSW	Rum	Econ	B	7	3	2	1	n	y	8	45	0	30
203	m	70	3	178	80	1	NSW	Beer	Econ	A	6	4	6	6	n	n	10	0	0	14
204	m	69	9	170	60	1	NSW	Malibu	Econ	B	7	5	5	2	y	n	8	365	0	30

ID	sex	yr	mon	ht	wt	ms	origin	drink	discip	gr	w	q	l	i	v	c	dist	pray	sm	TV
205	m	69	9	180	64	3	NSW	Gin	Law	C	7	6	7	2	y	n	6	0	0	20
206	f	69	4	178	67	*	NSW	Cooler	Accg	C	5	5	4	5	y	n	3	0	0	12
207	f	70	5	165	54	*	NSW	Tequila	Econ	B	4	3	7	7	y	n	7	0	0	10
208	m	70	1	170	63	1	NSW	Beer	Geog	B	5	2	6	3	n	n	7	1	0	10
209	f	66	7	*	*	1	NSW	Scotch	GEO	A	3	4	9	4	y	n	5	48	0	10
210	f	70	1	152	58	1	NSW	Cooler	Econ	A	9	0	4	0	y	y	30	30	0	20
211	f	70	4	165	56	1	NSW	Coke	Unsure	C	7	4	4	3	n	n	90	5	0	20
212	m	53	3	182	102	2	NSW	S.Comfort	Psych	B	4	8	2	7	y	y	12	48	0	15
213	m	62	9	185	80	1	NSW	Cider	Mass	C	4	4	2	2	y	n	0	1	0	4
214	m	69	8	178	60	1	NSW	Vodka	BHS	CQ	7	4	1	1	y	n	5	0	0	20
215	m	70	6	180	65	1	NSW	Port	Econ	CQ	4	3	0	0	y	y	30	1	0	30
216	m	69	9	180	75	1	NSW	Beer	Econ	A	9	2	0	0	y	n	35	0	11	24
217	f	69	7	175	61	3	NSW	Kalua	Econ	C	4	8	7	3	n	y	20	0	0	7
218	f	70	6	163	53	1	NSW	Lemon Sq	Mass	B	8	6	4	4	n	n	10	15	0	5
219	m	69	8	175	70	1	NSW	Beer	Sci	B	4	7	3	3	y	y	10	10	0	20
220	f	69	7	170	52	1	NSW	Kalua	Econ	B	6	3	9	5	n	n	15	0	0	6
221	m	69	11	*	65	1	NSW	Beer	Unsure	A	3	3	7	5	n	n	15	2	0	12
222	f	64	2	152	57	1	NSW	Scotch	BHS	B	6	3	6	3	y	n	10	15	10	8
223	f	49	8	162	65	1	NSW	Gin	Unsure	F	8	4	0	2	y	n	8	70	0	7
224	m	69	1	180	80	1	NSW	Beer	Unsure	C	8	5	7	3	y	n	5	0	5	15
225	f	63	*	157	57	2	NSW	Scotch	BHS	B	4	1	5	7	y	y	5	0	5	8
226	f	69	11	170	50	1	NSW	Malibu	Econ	B	8	5	8	5	n	n	30	25	0	21
227	f	70	4	160	47	1	NSW	Bacardi	Unsure	B	2	2	5	5	n	y	35	0	0	5
228	m	69	11	165	59	*	NSW	Heineken	Econ	A	7	2	7	0	y	y	13	0	0	4
229	f	70	6	167	50	3	NSW	Gin Lemon	Unsure	B	8	3	7	5	y	n	13	0	0	10
230	f	70	8	165	59	1	NSW	Fruit J	Unsure	C	7	2	2	0	n	n	20	25	0	15
231	m	70	5	177	75	1	NSW		Accg	B	4	4	6	1	y	n	25	4	0	15
232	f	70	4	165	57	1	NSW	Water	Accg	A	3	0	8	0	n	n	20	52	0	15
233	f	70	1	157	45	1	NSW	Iced Choc	Unsure	F	8	2	0	0	n	n	20	0	0	50
234	f	63	7	173	57	1	NSW	Champagne	Unsure	B	9	4	0	2	y	n	23	0	0	6
235	m	61	5	173	65	1	NSW	Beer	Comp	A	6	4	4	4	y	y	5	1	0	4
236	f	70	5	170	47	1	NSW	Bourbon	Accg	F	6	3	2	0	n	y	25	2	0	30
237	m	69	9	170	80	1	NSW	Mineral W	Econ	C	7	4	2	4	y	y	12	55	0	4
238	f	69	3	187	53	1	NSW	Gin	GEO	C	4	6	6	0	n	n	1	3	0	2
239	m	69	12	180	74	1	NSW	Bourbon	Econ	C	9	5	3	3	n	n	15	1	0	3
240	m	69	7	173	69	1	NSW	Bourbon	Econ	C	8	4	3	5	y	y	15	1	4	8
241	m	70	6	185	75	1	NSW	Bourbon	Accg	B	9	7	7	1	y	n	30	0	0	6
242	f	68	11	160	57	1	NSW	Lemon Sq	Chem	A	6	4	9	5	n	n	5	80	0	1
243	f	70	2	165	*	1	NSW	Scotch	Mass	C	7	3	3	0	n	n	14	0	0	30
244	m	70	2	180	65	1	NSW	Rum	Mass	B	8	5	8	4	n	y	14	40	0	4
245	f	69	9	170	55	1	NSW	Soft Dr	Comp	B	9	5	3	4	y	n	35	2	0	8

ID	sex	yr	mon	ht	wt	ms	origin	drink	discip	gr	w	q	l	i	v	c	dist	pray	sm	TV
246	m	70	3	175	71	1	NSW	Coke	Comp	B	7	6	5	4	n	n	20	60	0	8
247	m	70	1	170	100	*	NSW	Pimms	Comp	C	6	0	5	0	y	y	40	1	0	30
248	m	62	8	178	80	1	NSW	Beer	Unsure	C	7	4	2	3	y	n	10	4	0	2
249	m	69	2	*	66	1	NSW	Cooler	Unsure	CQ	9	1	1	0	y	n	30	0	0	2
250	f	70	2	175	58	1	NSW	Wine	Mass	B	6	6	4	4	n	n	0	0	0	1
251	f	68	8	175	68	1	NSW	Tia Maria	BHS	CQ	5	3	3	3	y	n	30	0	0	30
252	f	69	3	169	46	1	NSW	Tia Maria	Mass	C	4	2	1	1	y	n	17	0	0	0
253	f	67	8	158	52	1	NSW	Black Rus	Maths	C	4	3	3	2	y	n	60	10	0	10
254	f	69	2	160	54	1	NSW	Black Rus	Comp/M	B	2	1	6	4	n	n	25	20	0	1
255	m	67	9	175	60	1	NSW	Gin	Comp	B	4	4	5	4	n	n	0	0	0	0
256	f	69	6	155	50	1	NSW	Coke	Mass	C	8	2	2	0	n	y	40	52	0	20
257	m	69	10	178	75	1	NSW	Bourbon	Mass	C	6	3	1	4	y	y	17	2	3	25
258	m	69	12	160	60	1	NSW	Soft Dr	Comp	A	3	4	5	6	y	y	18	0	0	0
259	f	69	10	180	75	1	NSW	S.Comfort	Unsure	CQ	9	3	0	0	y	y	90	7	0	15
260	f	69	10	165	50	1	NSW	Tia Maria	Comp	C	6	5	5	1	n	y	10	10	5	12
261	m	69	12	190	95	1	NSW	Coopers	BHS	CQ	2	2	1	2	n	n	0	0	0	30
262	f	70	2	166	52	1	NSW	Scotch	Unsure	C	9	1	1	1	n	y	25	0	0	8
263	m	68	12	188	85	*	NSW	Coopers	BHS	CQ	4	4	4	0	n	n	15	0	25	10
264	m	68	12	158	61	1	NSW	Lemon Sq	Unsure	C	9	7	5	7	y	n	1	0	0	22
265	f	68	4	170	68	1	NSW	Beer	Unsure	C	5	2	4	5	n	n	25	0	20	6
266	m	70	4	*	75	1	NSW	Harvey W	Sci	B	9	4	5	3	n	n	15	50	0	7
267	f	70	3	165	55	1	NSW	Gin	Mass	A	1	6	9	3	y	n	20	0	0	10
268	f	69	10	172	59	1	NSW	Coke	Educ	B	7	2	2	4	y	n	30	60	0	14
269	f	69	3	168	*	1	NSW	S.Comfort	Educ	CQ	8	2	1	1	n	n	10	0	0	14
270	m	70	5	183	65	1	NSW	Vodka	Econ	B	4	4	9	0	n	n	20	0	0	14
271	f	69	11	170	55	1	NSW	Cointreau	Biol	CQ	4	7	3	1	n	n	5	1	0	4
272	f	69	9	172	54	1	NSW	Lemon Sq	Unsure	B	7	2	2	0	n	n	15	54	0	*
273	m	70	4	161	70	1	NSW	Beer	Unsure	C	8	5	2	0	y	n	10	7	10	14
274	m	68	12	173	85	1	NSW	Beer	Comp	B	4	9	9	8	y	n	18	40	0	20
275	f	70	4	*	*	1	NSW	S.Comfort	ENGL	C	6	8	7	4	n	n	35	0	0	20
276	f	69	11	158	57	1	NSW	Vodka	Unsure	A	9	7	8	6	y	n	30	35	0	13
277	f	70	4	164	53	1	NSW	S.Comfort	Mass	A	6	6	5	3	n	n	35	10	0	10
278	f	60	8	161	68	2	NSW	Gin	Unsure	B	6	8	7	4	y	n	10	0	0	6
279	f	67	1	158	60	1	NSW	Tequila	Biol	B	4	3	5	2	y	n	15	60	0	10
280	f	68	10	173	75	1	NSW	Gin	Geog	B	6	6	7	1	y	y	5	1	0	5
281	m	68	10	180	65	1	NSW	Vodka	Unsure	C	8	6	5	1	y	y	17	2	0	12
282	m	66	9	183	75	1	NSW		Econ	C	7	3	4	2	y	n	35	0	0	20
283	f	67	12	161	53	1	NSW		Educ	F	9	1	1	3	y	n	25	70	0	10
284	f	68	12	168	50	1	NSW	Gin	Mass	C	8	2	2	2	y	n	5	1	3	10
285	m	68	3	185	85	1	NSW	S.Comfort	Econ	B	8	8	7	3	n	y	10	5	0	40
286	m	48	2	162	52	2	NSW	Coke	Unsure	B	7	9	4	9	y	y	100	0	0	15

ID	sex	yr	mon	ht	wt	ms	origin	drink	discip	gr	w	q	l	i	v	c	dist	pray	sm	TV
287	f	68	4	168	62	1	NSW	Cointreau	Econ	B	9	6	7	7	n	y	0	0	0	0
288	f	68	4	172	70	1	NSW	Beer	Biol	CQ	9	1	4	4	y	n	0	0	0	1
289	f	68	9	170	65	1	NSW	Malibu	Psych	C	8	3	3	3	n	n	0	0	0	0
290	m	63	3	178	66	1	NSW	Beer	Biol	C	5	7	6	2	y	n	25	0	0	10
291	f	69	11	167	50	1	NSW	Lemon	Maths	C	9	4	0	4	y	y	5	20	0	10
292	f	70	4	164	67	1	NSW	Wine	Unsure	C	4	7	3	4	n	n	25	10	0	7
293	f	68	10	175	51	1	NSW	Bacardi	Accg	C	5	9	6	6	n	n	25	80	0	4
294	f	68	12	163	64	1	NSW	Coke	Maths	C	7	5	3	5	n	y	25	0	0	12
295	f	63	1	152	50	1	NSW	Lemon	Comp	C	6	6	4	7	n	n	5	10	0	5
296	m	57	3	187	79	3	NSW	Tonic	Comp	C	8	5	3	4	y	y	25	0	25	5
297	f	64	8	167	59	1	NSW	Coffee	Soc	CQ	6	6	1	1	y	n	20	0	0	0
298	m	69	8	*	*	1	NSW	Lemonade	Comp	C	6	3	8	7	n	y	4	1	0	5
299	f	68	2	152	38	1	NSW	Kalua	Econ	C	5	8	4	1	n	n	6	0	0	2
300	f	67	11	160	51	1	NSW	Beer	Unsure	C	4	7	4	0	n	n	20	0	25	10
301	f	68	12	168	60	3	NSW	Bacardi	Biol	B	4	6	5	5	n	n	15	0	0	15
302	f	70	1	169	60	1	NSW	Tia Maria	Psych	C	4	6	7	1	y	n	15	0	0	21
303	f	69	10	165	49	1	NSW	Coke	Maths	C	7	7	5	2	n	n	5	55	0	10
304	f	69	9	160	53	1	NSW	Tia Maria	Math/C	CQ	9	2	3	0	n		25	0	0	7
305	m	69	8	178	72	1	NSW	Vodka	Comp	C	7	2	4	1	y	n	25	5	0	6
306	m	69	9	180	82	1	NSW	Beer	Comp	C	9	6	6	0	n	y	20	60	0	20
307	m	68	8	170	56	1	NSW	Coke	Econ	A	7	3	4	1	n	n	0	52	0	2
308	f	68	11	157	49	1	NSW	Snowball	Comp	B	9	4	8	3	y	n	20	0	0	20
309	f	54	11	160	47	2	NSW	Scotch	Anthr	C	7	3	3	3	y	y	15	0	0	15
310	m	69	7	179	79	1	NSW	Coopers	Psych	C	7	7	6	5	y	n	5	0	15	10
311	m	61	10	175	70	*	NSW	Beer	Sci	B	4	4	4	4	n	y	25	0	25	4
312	m	64	4	180	65	1	NSW	Beer	Comp	B	6	6	6	5	y	n	15	0	0	8
313	m	65	3	180	70	1	NSW	Beer	Econ	C	2	8	6	8	n	y	15	0	25	10
314	f	69	11	158	50	1	NSW	Vodka	Psych	C	5	2	0	2	y	y	20	0	0	18
315	f	69	4	158	56	1	NSW	Lemon Sq	Accg	A	2	6	6	4	n	n	12	52	0	25
316	f	70	2	176	60	1	NSW	Daiquiri	Mass	B	9	4	3	0	n	n	0	0	0	0
317	f	67	9	173	57	1	NSW	Amanda	Biol	B	5	5	5	3	n	n	30	3	0	20
318	m	69	4	180	76	1	NSW	Beer	GEO	C	6	2	5	2	y	n	10	50	2	9
319	f	69	2	161	*	1	NSW	Bacardi		B	7	3	3	0	n	n	30	1	0	20
320	f	69	7	160	56	1	NSW	Beer	Biol	F	5	4	1	3	y	n	30	0	15	16
321	f	68	12	166	59	1	NSW	Champagne	BHS	C	4	4	2	2	n	y	12	64	0	5
322	f	70	1	150	55	1	NSW	Vodka	Psych	CQ	8	8	3	6	y		20	0	0	10
323	f	69	3	160	45	1	NSW	Tia Maria	Biol	B	7	8	5	1	y	y	30	52	0	10
324	f	50	11	150	49	2	NSW	Orange J	Psych	CQ	7	8	2	3	y	y	3	6	25	12
325	f	70	1	165	57	1	NSW	Soft Dr	Psych	A	7	3	5	5	n	n	15	60	0	10
326	f	48	11	160	58	2	NSW	Wine	Psych	A	9	8	4	4	y	y	8	0	0	4
327	m	69	10	185	75	1	NSW	Beer	Econ	B	9	8	4	1	n	y	15	0	0	8

ID	sex	yr	mon	ht	wt	ms	origin	drink	discip	gr	w	q	l	i	v	c	dist	pray	sm	TV
328	m	69	11	181	*	1	NSW	Bourbon	Hist	C	9	4	1	1	y	n	40	52	0	25
329	f	70	3	155	50	1	NSW	Brandy	Engl	CQ	8	3	5	1	n	n	10	5	0	5
330	f	48	7	130	75	2	NSW	Wine	Psych	C	5	5	2	4	n	n	*	0	0	12
331	f	41	7	120	45	3	NSW	Soda W	Pol	C	8	9	0	8	y	y	140	0	0	8
332	f	68	10	174	60	*	NSW	Cointreau	Law	A	5	5	7	4	n	y	18	52	0	4
333	f	70	5	168	65	1	NSW	Tequila	Educ	F	7	1	0	0	n	n	2	55	10	10
334	m	69	4	180	80	1	NSW	Beer	Biol	C	9	4	2	2	n	n	15	61	0	10
335	f	69	7	170	55	1	NSW	Soft Dr	Educ	CQ	4	4	0	0	n	y	25	0	0	5
336	f	69	9	163	55	1	NSW	Vodka	Accg	C	7	4	4	2	n	n	35	2	3	10
337	m	70	5	180	60	1	NSW	Cordial	Unsure	A	4	4	7	5	y	n	13	360	4	18
338	m	70	6	179	69	1	NSW	Orange J	BHS	C	5	3	2	2	n	n	5	40	0	10
339	f	69	10	*	*	1	NSW	Bourbon	Econ	C	4	2	2	0	n	n	*	52	0	10
340	m	69	12	170	65	1	NSW	Beer	Econ	B	6	1	1	0	n	n	20	0	0	15
341	m	69	6	171	57	1	NSW	Lemon Sq	GEO	C	2	4	4	1	n	n	4	60	0	14
342	m	70	2	175	75	*	NSW	Beer/Wine	Law	B	4	4	1	3	y	n	10	1	0	1
343	m	70	2	180	65	1	NSW		Econ	B	8	7	6	4	y	n	20	0	0	10
344	m	69	3	173	53	1	NSW	Beer	Econ	C	6	5	2	1	n	n	25	52	20	10
345	f	69	7	150	42	1	NSW	Wine	Econ	C	6	7	5	5	n	n	0	50	0	3
346	f	69	10	160	51	1	NSW	Scotch	Accg	B	3	3	3	0	n	n	1	1	10	7
347	f	70	4	165	50	*	NSW	Coffee	Unsure	CQ	0	9	9	0	y	y	30	60	0	23
348	f	68	7	178	62	1	NSW	Coke	Accg	C	4	4	7	3	n	n	6	0	0	2
349	f	69	4	150	41	*	NSW	Mineral W	Educ	C	8	4	4	3	n	y	25	50	0	2
350	f	69	9	170	64	1	NSW	Bacardi	Psych	C	6	4	2	1	n	y	20	60	0	5
351		70	3	172	57	1	NSW	Cointreau	Econ	B	8	5	3	1	y	n	25	20	0	6
352	f	70	3	172	56	1	NSW	Malibu	Pol	C	9	4	5	0	n	y	30	0	1	14
353	f	70	2	175	58	1	NSW	Tia Maria	Econ	C	4	3	3	1	n	n	25	*	0	5
354	f	70	4	160	47	1	NSW	Vodka	Econ	B	5	5	4	1	n	n	7	2	0	4
355	m	70	5	177	72	1	NSW	Vodka	Accg	B	4	7	4	2	y	y	10	2	3	15
356	m	70	3	180	57	1	NSW	Scotch	Unsure	C	8	0	3	0	n	y	5	0	0	14
357	m	70	3	178	80	1	NSW	Port	Acc/L	A	5	1	9	3	n	n	8	300	0	10
358	m	70	3	170	60	1	NSW	S.Comfort	Biol/L	B	8	2	6	3	y	y	18	365	0	12
359	f	69	5	169	58	1	NSW	Ouzo	Biol	F	9	4	0	0	n	n	0	0	25	1
360	f	70	2	170	65	1	NSW	Champagne	Accg	B	5	6	7	6	n	n	0	0	0	5
361	f	70	5	163	*	*	NSW	Scotch	Unsure	C	6	3	0	2	n	n	10	0	10	5
362	f	70	3	179	65	1	NSW	Beer	Mass	B	5	4	4	2	n	n	3	0	0	3
363	m	70	2	178	70	1	NSW	Bacardi	ECFS	B	8	6	7	4	n	y	20	1	0	20
364	m	69	12	174	68	1	NSW	Bourbon	Law	C	3	3	4	3	n	n	9	0	0	10
365	m	70	2	185	68	1	NSW	Beer	Accg	CQ	6	3	3	3	y	n	15	30	0	15
366	m	70	3	170	65	1	NSW		Unsure	B	1	5	6	4	y	n	30	1	0	10
367	f	70	7	170	62	1	NSW	Rum	Econ	C	4	7	7	4	n	n	15	2	0	10
368	f	68	10	165	84	1	NSW	Coke	Unsure	C	3	4	3	2	n	n	23	0	0	15

ID	sex	yr	mon	ht	wt	ms	origin	drink	discip	gr	w	q	l	i	v	c	dist	pray	sm	TV
369	f	70	3	163	43	1	NSW		Accg	B	6	4	4	2	n	n	0	3	3	2
370	f	68	6	160	60	1	NSW	Cider	Accg	B	7	3	3	3	y	n	20	0	20	20
371	f	69	1	174	55	1	NSW	S.Comfort	Mass	B	5	6	4	0	y	n	15	2	1	1
372	f	70	7	163	46	1	NSW		Psych	B	5	3	4	0	n	n	0	50	0	0
373	m	68	1	182	96	1	NSW	Beer	Econ	C	9	8	6	4	y	n	10	0	0	6
374	m	68	7	183	60	1	NSW	Lemon Sq	Psych	C	8	8	5	6	y	y	20	4	0	12
375	f	69	12	163	57	1	NSW	Rum	Psych	C	7	4	4	5	y	n	5	1	0	1
376	m	69	8	180	75	*	NSW	Coke	Econ	B	2	4	7	2	n	y	12	0	0	20
377	m	70	6	180	78	1	NSW	Heineken	Accg	A	0	3	9	3	n	n	12	30	0	10
378	f	70	1	174	54	1	NSW	Coke	Law	B	4	4	5	5	n	y	20	1	0	16
379	f	69	12	176	*	1	NSW	Coke	Econ	B	5	6	5	5	y	n	20	6	0	16
380	m	60	2	182	76	1	NSW	Beer	Law	A	4	4	9	7	n	y	20	0	0	1
381	f	48	4	172	65	3	NSW	Mineral W	Psych	A	8	6	6	6	y	n	12	40	0	0
382	m	70	6	*	52	1	NSW	S.Comfort	Econ	A	5	4	7	4	n	y	20	60	0	4
383	m	69	3	180	75	1	NSW	Beer	Psych	C	7	7	8	4	y	n	5	20	0	5
384	m	69	3	179	66	1	NSW	Muscat	Accg	A	3	7	7	7	n	y	25	60	0	9
385	m	69	6	177	89	1	NSW	Beer	Unsure	C	5	9	5	6	y	y	5	27	0	9
386	f	70	5	160	49	1	NSW	Water	Law	A	0	0	5	0	y		20	0	0	5
387	m	69	11	183	73	1	NSW	Beer	Accg	B	6	9	7	5	y	n	22	0	0	2
388	m	69	8	178	75	1	NSW	Black Rus	Econ	A	8	9	7	3	y	n	18	0	30	14
389	m	69	12	187	79	1	NSW	Tequila	Accg		6	4	4	3	n	n	15	0	15	12
390	f	69	12	170	60	1	NSW	Malibu	Unsure	CQ	8	2	2	1	n	n	60	2	0	14
391	f	68	12	169	58	1	NSW	Scotch	Mass	C	9	1	1	0	n	n	0	0	0	0
392	m	70	3	174	*	1	NSW		Accg	B	8	9	9	3	y	n	5	0	0	20
393	m	65	3	173	70	1	NSW	Orange J	Econ	B	8	8	5	6	y	n	30	75	0	20
394	m	68	11	181	65	1	NSW	Coke	Econ	C	3	4	3	7	y	n	15	54	0	5
395	m	70	4	185	70	1	NSW	Bourbon	Law	B	3	7	8	3	y	y	30	3	0	10
396	f	68	11	175	*	1	NSW	Wine	Mass	C	9	2	0	0	n	n	0	0	0	0
397	m	70	2	171	66	1	NSW	Scotch	Accg	B	9	4	4	1	n	n	10	25	0	25
398	f	68	6	174	51	1	NSW	Lemon Sq	Biol	B	2	5	4	2	n	n	18	0	0	2
399	f	70	4	168	62	1	NSW	Cointreau	Engl	C	7	4	2	0	n	y	0	1	8	2
400		69	10	185	80	1	NSW	Beer	Accg	C	7	5	6	3	y	n	10	10	0	*
401	m	69	11	188	80	1	NSW	Beer	Accg	C	6	4	3	3	n	y	10	8	0	15
402	f	70	7	160	45	3	NSW	Beer	Econ/L	C	8	5	3	0	y	n	6	1	6	10
403	f	69	11	158	50	1	NSW	Vodka	Unsure	C	6	5	3	0	y	y	5	0	0	12
404	f	70	2	170	60	1	NSW	Coke	Econ	C	8	4	5	1	y	n	20	0	0	10
405	f	70	6	157	60	1	NSW	Cooler	Accg	A	4	2	6	6	y	y	20	10	0	10
406	f	65	6	165	55	1	NSW	Mineral W	Accg	CQ	8	6	4	1	y	n	5	0	6	3
407	m	65	4	180	*	1	NSW	Beer	Econ	CQ	8	5	8	0	y	n	10	0	20	15
408	f	55	6	158	57	3	NSW	Rum	Psych	B	5	7	5	6	y	y	3	0	0	20
409	m	70	3	176	66	1	NSW	Kalua	Econ	B	5	3	5	3	n	n	20	52	0	2

ID	sex	yr	mon	ht	wt	ms	origin	drink	discip	gr	w	q	l	i	v	c	dist	pray	sm	TV
410	m	69	10	*	67	1	NSW	Bacardi	Acc/L	B	4	4	5	3	n	n	0	1	0	1
411	f	69	9	175	50	1	NSW	Coke	Pol	B	5	6	5	4	n	n	0	70	0	1
412	f	70	5	170	52	1	NSW	Coke	Mass	B	5	5	4	0	n	y	0	70	0	1
413	f	69	12	175	59	1	NSW	Kalua	Psych	C	8	5	2	3	y	y	20	0	0	10
414	m	67	1	180	70	2	NSW	Beer	GEO	B	6	3	5	2	y	n	30	0	4	15
415	m	70	7	180	69	1	NSW	Beer	Mass	A	5	6	9	1	y	y	50	0	0	44
416	m	69	11	178	62	1	NSW	Scotch	Econ	C	8	3	4	1	y	n	15	0	0	*
417	f	69	9	170	75	1	NSW	Malibu	Accg	C	6	7	7	7	y	n	*	10	8	10
418	f	70	3	170	50	1	NSW	Kalua	Mass	B	6	8	2	2	n	n	10	10	0	5
419	m	69	9	*	75	1	NSW	Kalua	Unsure	C	9	1	0	4	y	n	25	1	0	5
420	f	67	11	150	46	1	NSW	Bourbon	Psych	B	5	6	7	5	y	n	20	0	0	14
421	m	70	4	175	60	1	NSW	Wine	Maths	C	6	5	4	3	y	y	30	10	0	30
422	f	69	9	170	75	1	NSW	Gin	Engl	CQ	9	2	1	1	n	y	35	60	0	8
423	f	70	6	170	*	1	NSW	Bacardi	Mass	C	2	5	4	2	n	n	0	30	0	4
424	f	69	9	151	51	1	NSW	Bacardi	Accg	B	6	3	6	3	n	n	0	0	0	16
425	f	69	9	*	*	1	NSW	Kalua	Educ	C	6	4	4	3	n	n	0	30	0	5
426	f	70	1	160	52	1	NSW	S.Comfort	Mass	C	7	2	2	0	y	y	97	40	0	10

References

Atwell, B.J. (1991): Factors which affect the growth of grain legumes on a solonised brown soil. I: Genotypic responses to soil physical factors, *Australian Journal of Agricultural Research*, **42**, pages 95-105.

Barnicot, N.A. & D.R. Brothwell (1959): The evaluation of metrical data in the comparison of ancient and modern bones, in *Medical Biology and Etruscan Origins*, G.E.W. Wostenholme & C.M O'Connor (eds), Little, Brown & Co., page 136.

Brown, Jennifer & Jane Fielding (1993): Qualitative differences in men and women police officers' experience of occupational stress, *Work and Stress*, **7**(4), pages 327-340.

Cotton, S., J.K. Antill & J.D. Cunningham (1990): The work attachment of mothers with preschool children, *Psychology of Women Quarterly*, **14**, pages 255-270.

Cox, D.R. & E.J. Snell (1981): *Applied Statistics*. London: Chapman & Hall.

Dale, G., J.A. Fleetwood, A. Weddell, R.D. Ellis & J.R.C. Sainsbury (1987): Beta endorphin: a factor in 'fun run' collapse? *British Medical Journal*, **294**, page 1004.

Donovan, J.W., R. MacLennan & M. Adena (1984): Vietnam service and the risk of congenital anomalies: a case-control study, *Medical Journal of Australia*, **140**, pages 394-397.

Floud, R. (1973): *An Introduction to Quantitative Methods for Historians* (2nd ed.). London: Methuen.

Gastwirth, J. L. (1988): *Statistical Reasoning in Law and Public Policy, vol 2: Tort Law, Evidence and Health*. New York: Academic Press.

Gilchrist, W, (1984): *Statistical Modelling*. Chichester: John Wiley & Sons.

Grizzle, J.E., C.F. Starmer & G.G. Koch (1969): Analysis of categorical data by linear models, *Biometrics*, **25**, pages 137-156.

Hampton, J.R. & J.D. Hill (1978): A randomised trial of home-versus-hospital management for patients with suspected myocardial infarction, *The Lancet,* 22 April, pages 837-841.

Hand, D.J., F. Daly, A.D. Lunn, K.J. McConway & E. Ostrowski (1994): *A Handbook of Small Data Sets.* London: Chapman & Hall.

Harper, J.F. & E. Marshall (1991): Adolescents' problems and their relationship to self-esteem, *Adolescence*, **16** (104), pages 799-808.

Henderson-Sellers, A. (1990): Australian public perception of the greenhouse issue, *Climatic Change*, **17**, pages 69-96.

Higham, C.W.F., A. Kijngam & B.F.J. Manly (1980): An analysis of prehistoric canid remains from Thailand, *Journal of Archaeological Science*, **7**, pages 148-165.

Johnson, O. (1991): *Information Please Almanac*. Boston: Houghton-Mifflin.

MacArthur, J.W. (1931): Linkage studies with the tomato. III: Fifteen factors in six groups, *Transactions of the Royal Canadian Institute*, **18**, pages 1-19.

Mazess, R.B., W.W. Peppler & M. Gibbons (1984): Total body composition by dual-photon (^{153}Gd) absorptiometry, *American Journal of Clinical Nutrition*, **40**, pages 834-839.

McGregor, G.A., N.D. Markandu, J.E. Roulston & J.C. Jones (1979): Essential hypertension: effect of an oral inhibitor of angiotensin-converting enzyme, *British Medical Journal*, **2**, pages 1106-1109.

McNemar, Q. (1947): Note on the sampling error of the difference between correlated proportions or percentages, *Psychometrica*, **12**, pages 153-157.

Mendenhall, W., W. Ott & R.F. Larson (1974): *Statistics: A Tool for the Social Sciences*. Boston: Duxbury Press.

Pagano, M. & K. Gauvreau (1993): *Principles of Biostatistics*. Belmont, CA: Duxbury Press.

Pocock, S. (1983): *Clinical Trials*. Chichester: John Wiley & Sons.

Rousseeuw, P.J. & A.M. Leroy (2005): *Robust Regression and Outlier Detection.* New York: John Wiley & Sons.

Selvin, S. (2004): *Statistical Analysis of Epidemiological Data.* New York: Oxford University Press.

Sleigh, A., R. Hoff, K. Mott, M. Banneto, T. Maisk de Paiva, J. de Sousa Pedrosa & I. Sherlock (1982): Comparison of filtration staining (Bell) and thick smear (Kato) for the detection and quantitation of *Schistosoma mansoni* eggs in faeces, *Transactions of the Royal Society of Tropical Medicine and Hygiene,* **76**, pages 403-406.

Sokal, R.R. & F.J. Rohlf (1981): *Biometry*, (2nd edition). San Francisco: W.H. Freeman.

Solowij, N., P.T. Mitchie & A.M. Fox (1991): Effects of long-term cannabis use on selective attention: an event-related potential study, *Pharmacology Biochemistry & Behavior*, **40**, pages 683-688.

Statistical Abstract of the U.S. (1988): Department of Commerce, USA.

Sternberg, D.E., D.P. Van Kammen & W.E. Bunney (1982): Schizophrenia: dopamine *b*-hydroxylase activity and treatment response, *Science*, **216**, pages 1423-1425.

Tan, K. & F. Choo (1990): A note on the academic performance of deep-elaborative versus shallow-reiterative information processing students, *Accounting and Finance*, May, pages 67-77.

Till, R. (1974): *Statistical Methods for he Earth Scientist*. London: McMillan.

Tukey, John (1986): Sunset salvo, *The American Statistician*, **40**, pages 72-76.

UNICEF (1992): *The State of the World's Children. Part 2: Statistical Tables*. The United Nations.

Utts, Jessica M. (1996): *Seeing Through Statistics.* Belmont, CA: Wadsworth.

Warton, P.M. & J.J. Goodnow (1991): The nature of responsibility: children's understanding of "your job", *Child Development*. **62**, pages 156-165.

SOLUTIONS

Chapter 2: Graphing Data

2.1

```
SO2 content   N  = 41
Leaf Unit = 1.0

 3    0    899
 18   1    000011234446778
(9)   2    346688999
 14   3    01156
 9    4    67
 7    5    66
 5    6    159
 2    7
 2    8
 2    9    4
 1    10
 1    11   0
```

Most levels are less than 70. The distribution is skewed to the right.

2.2

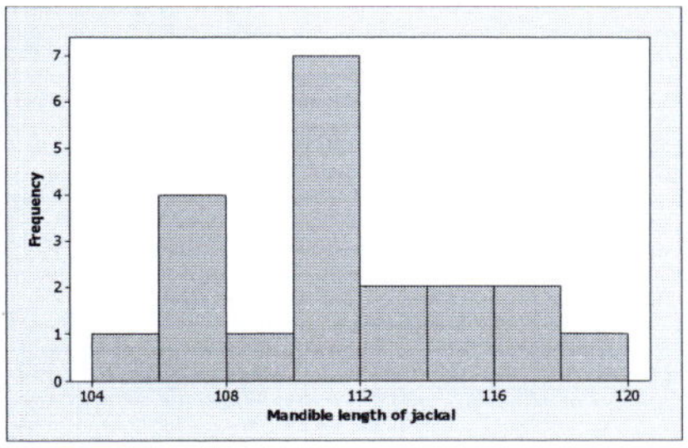

2.3

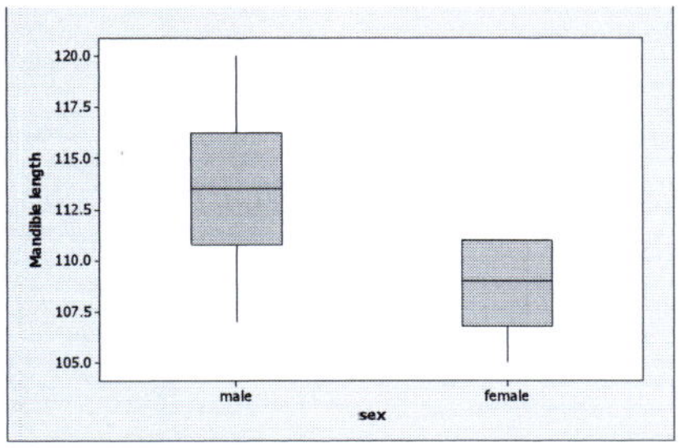

2.4

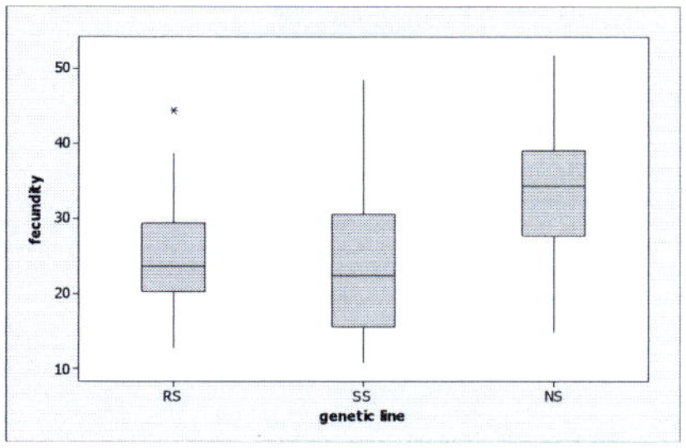

Median values are similar for species RS and SS whilst the median value is highest for species NS. Species SS has the largest midspread. Species RS has a high outlier.

2.5

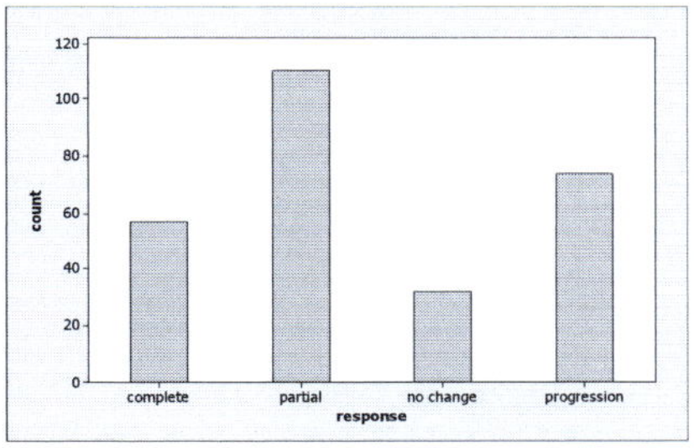

2.6

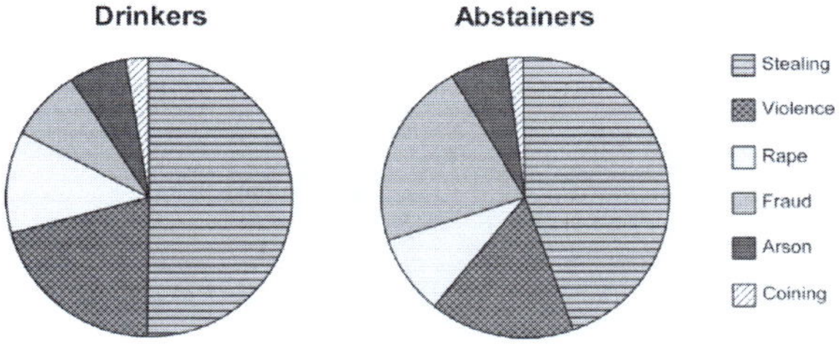

2.7b

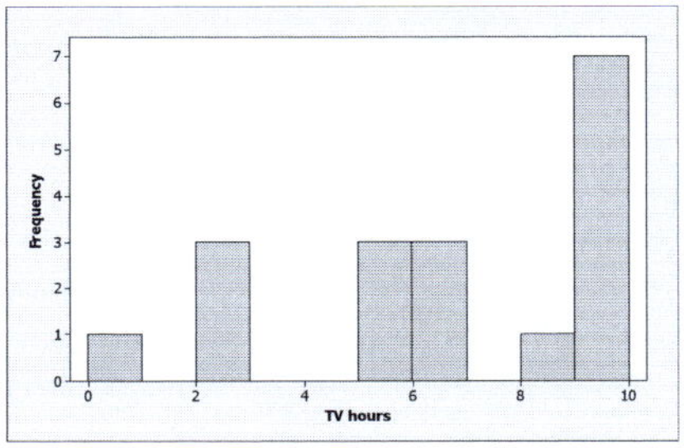

2.7c

```
Weights of women  N  = 21
Leaf Unit = 1.0

 1   4  1
 4   4  578
 10  5  002344
(5)  5  57779
 6   6  0000
 2   6
 2   7  0
 1   7
 1   8  4
```

This sampling procedure is valid provided the starting point is randomly allocated.

2.8a

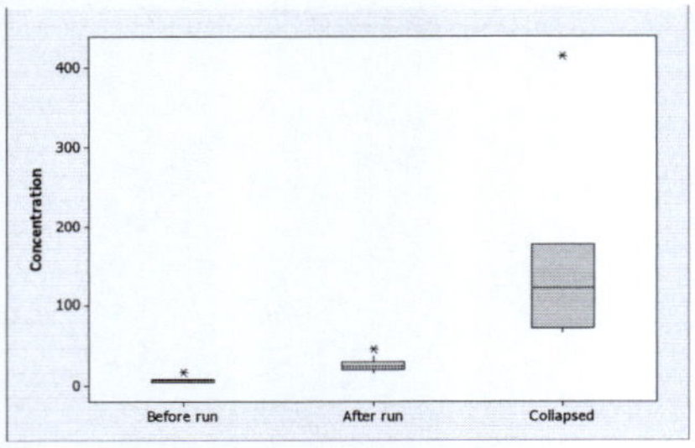

2.8b

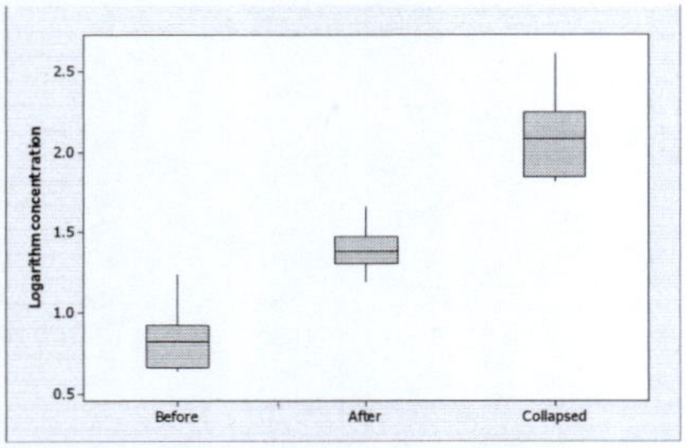

Box plots are difficult to interpret using the original data but taking logarithms to base 10 overcomes this problem. The medians of the 3 groups are quite different with the median being highest for the runners who collapsed. The variations are similar for the 3 groups.

2.9

```
Salary  N  = 14
Leaf Unit = 1.0

 2   3  05
(6)  4  225555
 6   5  0
 5   6  7
 4   7  007
 1   8  5
```

2.10

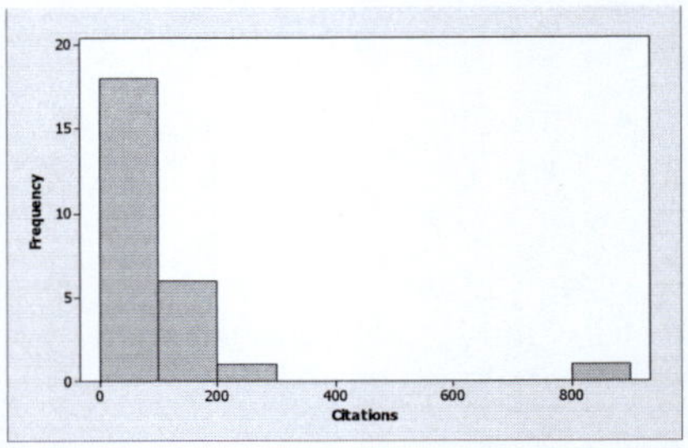

From the histogram it can be seen that the number of citations is heavily skewed to the right. This is due to H.J. **Eysenck** who had an unusually large number of citations (813).

Chapter 3: Summarising Data

3.1a **Sample 1**: 11/18
Sample 2: 9/18

3.1b **Sample 1**: Mean = 174.5 sd = 9.14
Sample 2: Mean = 164.72 sd = 15.63

3.1c **Sample 1**: Mean = 7.67 sd = 8.02
Sample 2: Mean = 6.412 sd = 4.109

3.1d **Sample 1**: Median = 4.50 IQR = 9.5
Sample 2: Median = 6.00 IQR = 5.5

Looking at a stem-and-leaf plot the distribution appears to be skewed to the right for the first sample so the median is more appropriate. Either the mean or median may be used as a summary statistic for the second sample since the distribution is approximately symmetric.

3.2a Median = 26, IQR = 22

3.2b Mean = 30.05 sd = 23.47

3.3

Type	Min	LQ	Med	UQ	Max
RS	12.8	20.3	23.6	29.3	44.40
SS	10.8	16.0	22.5	30.2	48.50
NS	14.9	28.2	34.4	37.9	51.80

3.4

Sample	Geometric Mean	Median
Before	7.63	7.20
After	25.6375	25.00
Collapsed	119.730	107.60

All three distributions were skewed but the distributions of the transformed data were not. If the arithmetic mean is a suitable measure of the centre of a data set that has been transformed by logarithms, then the geometric mean is a suitable measure of centre of the original data set. Here the geometric means are not far from the sample medians so we can conclude that the geometric means are suitable measures of centre for the original data sets.

3.5a n = 26, 20th %ile at 6th obs. = 68

3.5b n = 26, 95th %ile at 25th obs. = 251

3.6

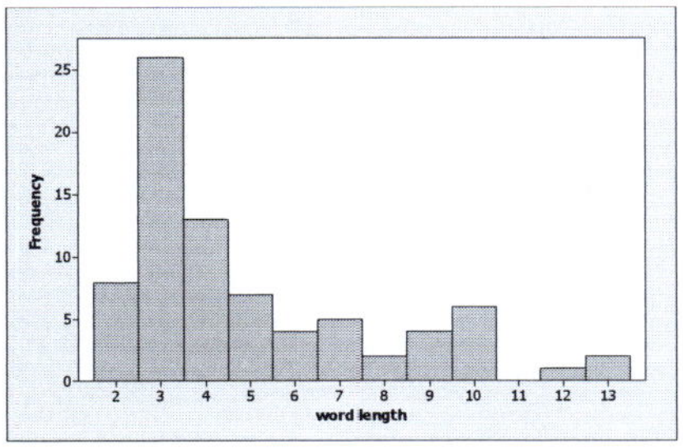

3.6a median = 4, LQ = 3, UQ=7, range=11, IQR = 4

3.6b mean = 5.026, sd = 2.857

3.6c Since the data are skewed then the median and quartiles are better measures of location and spread.

3.6d It is one way of determining possible authorship of an article.

3.7a Mean is closer to 5 than to 1 or 10. Data are approximately symmetrical so mean value is (roughly) the value in the middle of the range ie 5. A rough estimate of the standard deviation can be made by dividing the range by 4. Thus the estimate of the standard deviation is closer to 3 than to 6.

3.7b If 5 is added to each value, the mean will increase by 5 but the standard deviation will remain unchanged.
If each value is multiplied by 13, both the mean and the standard deviation will also be multiplied by 13.
If the sign is reversed on each value the sign on the mean will be reversed but the standard deviation will remain positive.

3.7c Mean = 5.211, median = 5, range = 10, sd = 3.102

3.8

State	Sample Size	Mean	Standard Deviation
ACT	3	6.000	0.00
NSW	27	6.296	1.96
QLD	8	5.625	2.066
S.A.	5	6.200	3.110
VIC	5	6.200	1.789
W.A.	3	6.667	1.528

3.9 n = 51,
10th percentile at 6th obs = 0
90th percentile at 46th obs. =25.

3.10 In the first 186 responses, 10 students failed to record a drink preference so these records will be ignored. Of the remaining 176 responses the 5 most popular drinks are:

Drink	Frequency	Percentage (from 176 Responses)
Beer	22	12.5
Coke	15	8.52
Wine	14	7.95
OJ	13	7.39
Soft Drink	8	4.55

Chapter 4: The Normal Distribution

Note: When answering questions involving areas under the normal curve, you should sketch the distribution and mark the relevant area.

4.1a y = 175: z = 3.12
Tab area = prob. = 0.00090

4.1b y = 165: z =1.88
Tabulated area = 0.0301
Prob. = 1 – 0.0301 = 0.9699

4.1c y = 135: z = -1.88
Tab area = prob. = 0.0301.
Customers complain about 3% of the time.

4.2a z = 1.04 at the 85th percentile.
y = 48120
$48,120 cuts off the top 15%.

4.2b z = 0.67 at the 75th percentile.
y = 47010
Since Werner's salary, $41500 is less than $47010, he doesn't get a key.

4.2c z = -1.28 at the 10th percentile.
y = 41160
Those earning less than $41,160 will be let go.

4.2d z = 0.84 at the 80th percentile.
y = 47520
Those earning at least $47,520 eat at *Madame's*.
z = -0.52 at the 30th percentile.
y = 43440. Those earning up to $43,440 bring their own lunches.
Those earning between $43,440 and $47,520 eat at the cafeteria.

4.3a y = 100: z = - 0.4
Tabulated area = probability = 0.3446.
Expected number sent back:
$0.3446 \times 130 \cong 45$.

4.3b From (a), the probability of an IQ less than 100 is 0.3446.
y = 110: z = 0.60
Tabulated area = probability = 0.2743
Probability of an IQ between 100 and 110 is
1 – (0.3446 + 0.2743) = 0.3811.
Expected number to join the army:
$0.3811 \times 130 \approx 49$

4.3c Prob. of an IQ>110 is 0.2743
Expected number to go to university:
0.2743*130 ≈ 36

4.4 *Verry ripe* bars may be rejected for being too heavy or too light. If these events are equally likely to occur, then 0.75% will be rejected because they are too light and 0.75% will be rejected because they are too heavy.
z = -2.43 cuts off an area of 0.0075 in the lower tail.
If z = -2.43: y = 38.06
z = 2.43 cuts off an area of 0.0075 in the upper tail.
z = 2.43: y = 41.94
Any *verry ripe* bar weighing < 38.06 or > 41.94 grams will be rejected.

4.5 Proportion = 0.0444

4.7a y = 56.7: z = -1.75
Tabulated area = 0.0401
y = 58.5: z = 2.75
Tabulated area = 0.0030
Probability = 0.0401 + 0.0030 = 0.0431
Expected number to be rejected:
$1000 \times 0.0432 \approx 43$

4.7b y = 56.8: z = -1.50
Tabulated area = 0.0668
y = 57.2: z = - 0.50
Tabulated area = 0.3085
Probability = 0.3085 - 0.0668 = 0.2417
Expected number to be rejected:
$1000 \times 0.2417 \approx 242$

4.7c z = 0.84 at the 80th percentile.
y = 145.52cm.

4.8a Assuming normality
y = 1.2: z = 0.8
Tabulated area = proportion = 0.2119

4.8b y = 0.8: z = 0.8
Tabulated area = 0.2119
y = 1.05: z = 0.2
Tabulated area = 0.4207
Prob. = 1 – (0.2119 + 0.4207) = 0.3674 ie. 36.74%

4.8c From b. 21.19% of worms eat less than 0.8g per day.
Expected no. $21.19\% \times 1000 \approx 212$

4.8d z = 1.28 at the 90th percentile.
y = 1.32g.

4.9

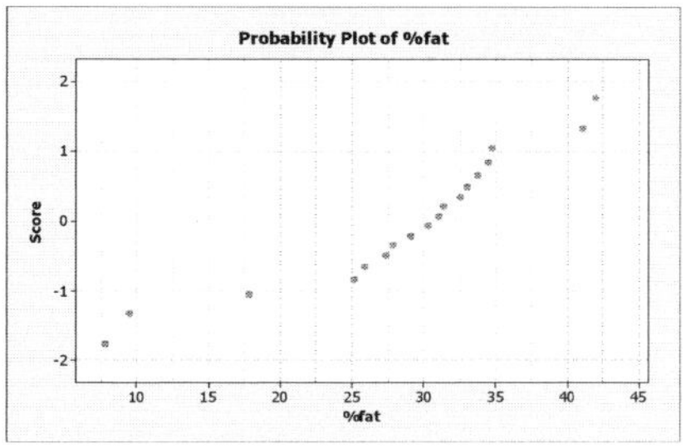

Data does not appear normal

Chapter 5: Distributions of Means and Proportions

5.1a $\bar{y}$ = 58: z = 2.60
Tabulated area = probability = 0.0047.

5.1b $\bar{y}$ = 141: z = - 5.16
Tabulated area = probability $\cong 0$.
5.1c Since these tennis balls have a much lower average bounce than is required, the club should complain.

5.2 Let π be the probability of rain.
$\pi = 0.25$, p = 9/17 = 0.529.
Now $n\pi = 17 \times 0.25 = 4.25$ and $n(1-\pi) = 17 \times 0.75 = 12.75$. Since $n\pi < 5$, the Central Limit Theorem for proportions does not apply and the normal approximation is not valid. Hence the results may be dubious.
p = 0.529: z = 2.66
Tabulated area = probability = 0.0039

5.3a $\bar{y}$ = 22.13: z = 4.26
Tab. area = probability < 0.000032

5.3b $\bar{y}$ = 21.63: z = 3.16
Tab. area = probability = 0.00079
The original suggestion that it takes 20 minutes, on average, to travel this distance could be incorrect.

5.4a $\bar{y}$ = 1100/1000 = 1.1g per worm per day, on average.

$\bar{y}$ = 1.1: z = 12.65

Tabulated area = probability ≅ 0
ie. It is highly unlikely

5.4b If 1,000 worms eat 6.5kg in total in 7 days then y-bar = 6500/(1000 x 7) = 0.93g eaten per worm per day.

$\bar{y}$ = 0.93: z = -23.43

Tabulated area ≅ 0, so probability ≅ 1
It is highly likely that 1,000 worms can eat 6.5kg in 7 days.

5.5a $\pi = 0.20$
Expected number = 0.2 × 80 = 16 children.

5.5b p = 50 / 200 = 0.25, n=200
Now since $n\pi$ and $n(1-\pi)$ are both ≥ 5,
CLT for proportions holds
p = 0.25: z = 1.77
Tabulated area = probability = 0.0384

5.6 $\pi = 0.166$, p = 1000/5000 = 0.2:
Check $n\pi$ and $n(1-\pi) \geq 5$, z = 6.46
Tabulated area = required area ≈ 0.

5.7 Let p be the sample proportion of graduate students needing a computer at any given time. The total number of graduate students, n, is 22. The average number of computers needed is 11 (50% of 22). Thus p, the sample proportion, has a distribution with a mean π = 0.5 and SE = 0.1066 (ie $\sqrt{\pi(1-\pi/n}$). In addition, since $n\pi$ = 22 ×0.5 = 11 and $n(1-\pi)$ = 22 ×0.5 = 11 are ≥ 5, the CLT will apply ie the distribution of the sample proportions will be approximately normal.
If N is the number of computers in the laboratory then there will be enough computers provided p is less than N/22. We need to find N such that the probability that p is less than N/22 is 0.9. From the standard normal tables we find that the z-value corresponding to an upper tail area of 0.1 is 1.28.
Now $p = \pi + (\sqrt{\{(\pi(1-\pi))/n\}}) \times z$
$= 0.5 + 0.1066 \times 1.28$
$= 0.636$
The proportion of computers to students (N/22) needs to exceed 0.636, so N needs to exceed 13.992 (0.636 × 22) ie 14 computers must be provided.

5.8a Assume normality.
y = 2.5: z = 0.5
Tabulated area = 0.3085.
Probability = 1 – 0.3085 = 0.6915

5.8b Assume normality, since n is small

$\bar{y}$ = 2.5: z = 1.0

Tabulated area = 0.1587.
Probability = 1 – 0.1587 = 0.8413

5.8c Prob. y > 2.5 = 0.6915 (from a)
If the four students are independent of each other, the probability that all four students will pass = 0.6915^4 = 0.2286

5.8d Since 0.8413 > 0.2286, there is a greater chance of passing if in a team.

5.8e Since the standard error of a mean decreases as n increases, a larger team will have a higher probability of passing so lowering the average time required seems reasonable.

5.9a y = 22: z = - 2.6
Tabulated area = proportion = 0.0047

5.9b $\bar{y}$ = 30: z = - 3.46
Tabulated area = probability = 0.00027

5.9c z ≈ 0.25 at the 60th percentile.

$y = 36.25g$

5.9d Since the heaviest 40% of the strawberries are no longer available, only 60% of the original distribution remains. The median will be at the centre of the remaining 60% ie. at the 30th percentile of the original distribution, where $z \approx -0.52$.

$\tilde{y} = 35 + (5 \times -0.52) = 32.4g$.

5.9e It is assumed that one punnet has been selected at random and that the shipment will be rejected if the mean weight of the strawberries is less than 30 grams.

$\bar{y} = 30$: $z = -2.11$

probability = 0.0174

5.9f Since the probability is only 1.74%, Ray should not worry.

Chapter 6: Confidence Intervals

6.1 $p = 120/500 = 0.24$
Check np and $n(1-p) \geq 5$
95% CI for π is (0.203, 0.277)

6.2a $p = 4/41000 = 0.0000976$
$np = 4$ and $n(1-p) = 40996$.
Since $np < 5$, the CLT for proportions will not apply ie. sample proportions will not follow a normal distribution. Hence it is not appropriate to provide a confidence interval.

6.2b No. Although the chance of a professional golfer hitting a hole-in-one with any one shot is given as ≈ 0.0001 it is not unusual for a professional golfer somewhere at some time to hit a hole-in-one.
Thus a hole-in-one is not usually news-worthy. The only reason that this report was considered news-worthy was that two professional golfers hit two consecutive holes-in-one.
The two outcomes can be considered to be independent of one another so the second outcome is not influenced by the first.
Given the occurrence of the first hole-in-one, the probability of the second is thus $4/41000 \approx 0.0001$, which is a lot less than one in 100 million!

6.3a $\bar{y} = 111.429$, $\sigma = 14$, $n = 14$
A 95% CI for μ is (104.1, 118.8)

6.3b p-value for test of normality = 0.937
So normality assumption is OK.

6.4 Normality check ($p = 0.885$)
$\bar{y}=0.0243$, $s=0.00514$, $n=10$, $t^*_9 =2.262$
A 95% CI for μ is (0.021, 0.028)

6.5 Normality check ($p = 0.011$)
Treat results with caution.
$\bar{y} =46.333$, $s =13.218$, $n =18$, $t^*_{17} = 2.11$
A 95% CI for μ is (39.8, 52.9)

6.6 Normality check ($p = 0.654$)
$\bar{y}=0.0164$, $s=0.00470$, $n=15$, $t^*_{14} =2.145$
A 95% CI for μ is (0.014, 0.019)

6.8a Normality check ($p = 0$)
$\bar{y}=891.16$, $s=967.3$, $n=25$, $t^*_{24} =2.064$
A 95% CI for μ is (491.9, 1290.5)

6.8b The assumption of normality has been violated ($p = 0$). However, after taking

logs the p-value for normality is 0.232.
Using the logarithms of the tonnages:
$\bar{y}$ =2.594, s =0.659, t^*_{24} = 2.064
A 95% CI for log μ is (2.322, 2.966
Taking antilogarithms gives a 95% CI for μ as (209.9, 734.5)

6.9 $\sigma = 0.4\% = 0.004$, $\delta = 0.1\% = 0.001$
Note: δ is half the width of the confidence interval.

$$\delta = 1.96\times(\sigma/\sqrt{n}) \quad \therefore n = \left(1.96\times\frac{\sigma}{\delta}\right)^2 = 61.5$$

Approximately 62 people are needed.

6.10 $\pi = 0.1$, $\delta = 0.01$
Note: δ is half the width of the confidence interval.

$$\delta = 1.96\times\sqrt{\pi(1-\pi)/n}$$

$$\therefore 0.01 = 1.96\times\sqrt{0.1\times 0.9/n}$$

$$\therefore n = 3457.4$$

Approximately 3457 responses are needed.

Chapter 7: Hypothesis Testing

7.1 H_0: $\pi = 0.01$
$n\pi$ and $n(1-\pi) \geq 5$. So, CLT applies.
z = 2.48, p = 0.0132.

7.2a H_0: $\pi = 0.113$
z = - 7.31, p-value = 0.

7.2b H_0: $\pi = 0.113$
z = 4.66, p-value ≈ 0.

7.2c **For part a**: $n\pi$ =51, $n(1-\pi)$ = 403
The normality assumption is met.
For part b: $n\pi$ = 1.9 and $n(1-\pi)$ = 15.1. The normality assumptions are not met.

7.3 H_0: $\pi = 0.01$
$n\pi$ and $n(1-\pi) \geq 5$. So, CLT applies.
z = 2.61, p-value = 0.0090.

7.4 H_0: $\mu = 60$
$\bar{y}$ = 67, s = 16, n = 43
The sample size is large enough for the CLT to apply ie. sample means will have an approximately normal distribution.
$t_{42} = 2.87$, $0.005 < \text{p-value} < 0.01$.

7.5 H_0: $\mu = 0.5$
Normality check (p=0.014)
$t_{26} = -0.32$, p-value > 0.5
Normality assumption may be violated.

7.6 H_0: $\mu = 9$
Normality check (p=0.0069)
$t_{15} = -1.85$, $0.05 < \text{p-value} < 0.1$
Due to the doubt about the assumption of normality of the parent population the results must be regarded with caution.

7.7 H_0: $\mu = 540$
z = 1.24, p-value = 0.2150.

7.8 H_0: $\mu = 142$
Normality check (p=0.372)
$t_{83} = 2.73$, $0.005 < \text{p-value} < 0.01$

7.9 H_0: $\pi = 0.001$
$n\pi = 1.7$ and $n(1-\pi) = 1693$.
Since $n\pi < 5$, the normality assumption has been violated. Thus the results should be regarded with caution.
z = 3.26, p-value = 0.00112.

7.10 The theory for this question is covered in Chapter 8: Comparisons
H_0: $\mu_d = 0$
Normality check of differences (p=0.91)
$\bar{d}$ = 4.073, s_d = 3.858, n = 30
$t_{29} = 5.78$, p-value < 0.0005.

Chapter 8: Comparisons

8.1a

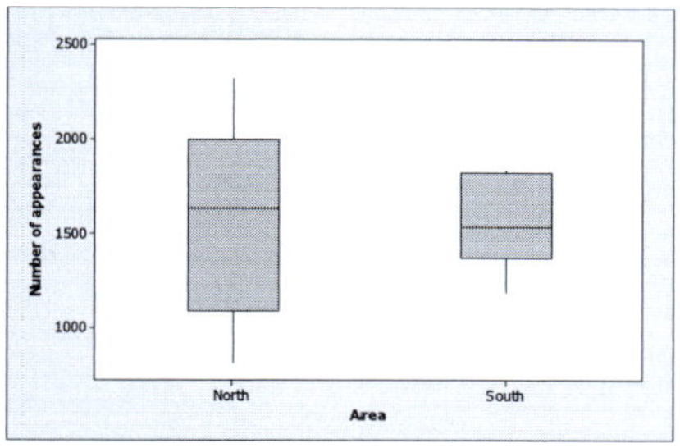

The median number of appearances in court per 100,000 population is slightly higher in the northern suburbs than in the southern suburbs. The spread for the northern suburbs appears higher than that from the southern suburbs. There are no outliers in either group.

8.1b H_0: $\mu_N = \mu_S$ or H_0: $\mu_N - \mu_S = 0$
$s_p = 435$
(check s_p lies between s_N and s_S)
Check assumptions of normality and equal spreads.
$t_{11} = 0.07$, p-value > 0.5.

8.2a Proportion of male drinkers = 0.191
Proportion of female drinkers = 0.131

8.2b H_0: $\pi_m = \pi_f$.
Check all np and $n(1\text{-}p) \geq 5$
z = 1.51, p-value = 0.1315.

8.2c A 95% CI for π_m is (0.153, 0.229)
A 95% CI for π_f is (0.0711, 0.191)

8.3a H_0: $\mu_D = 0$
Normality check (p=0.776)
$\bar{d} = -3.54$, $s_d = 15.092$ n = 13
$t_{12} = -0.85$, $0.2 < p < 0.5$.

8.3b H_0: $\mu_D = 0$
Normality check (p=0.438)
$\bar{d} = 1.615$, $s_d = 12.803$, n = 13
$t_{12} = 0.45$, p-value > 0.5.

8.3c

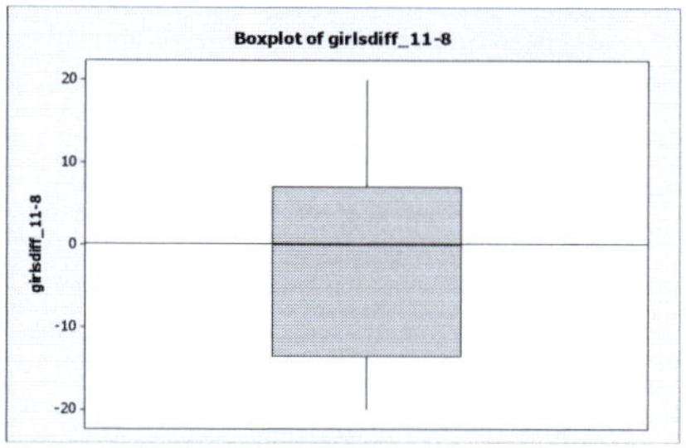

The box plot supports the conclusion reached in (b), that the average difference could be zero.

8.4 H_0: $\pi_1 - \pi_2 = 0$
z = 2.27, p-value = 0.0232

8.5a
MA students:
H_0: $\mu_1 = \mu_2$
$t_{46} = 8.67$, p-value < 0.0005
IS students:
H_0: $\mu_1 = \mu_2$
$t_{39} = 6.41$, p-value < 0.0005

8.5b H_0: $\mu_M = \mu_{IS}$
$t_{30} = 0.92$, p-value > 0.2

8.5c The assumption that the two populations have the same spread is dubious in the second test so a two-sample unpooled t-test is appropriate.

8.6

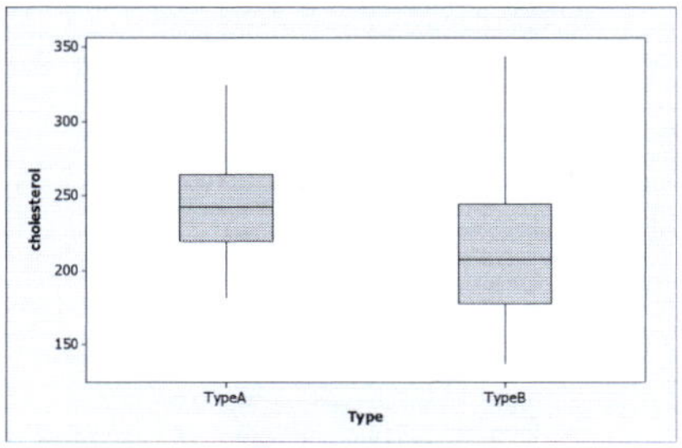

H_0: $\mu_A = \mu_B$
Check assumptions of normality and equal spreads.
$t_{38} = 2.58$, $0.01 < p < 0.02$

8.7 $H_0: \mu_m = \mu_p$
Check assumptions of normality and equal spreads.
$s_p = 0.3234$
$t_{40} = -4.07$, p-value < 0.0005

8.9a Differences: unemployment rates

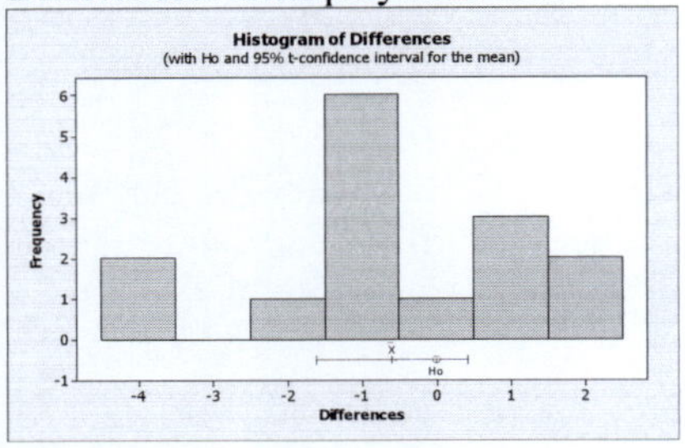

8.9b $H_0: \mu_D = 0$
Normality assumption OK from plot.
$\bar{d} = -0.60$, $s_d = 1.844$, $n = 15$
$t_{14} = -1.26$, $0.2 <$ p-value < 0.5

8.10 Since only methods covered in Chapter 8 are to be used in the analysis, areas will be compared two at a time.
Note: graphs of each of the samples are needed to check the assumptions of normality and equal stdev.

$\bar{y}_1 = 37.313$ $s_1 = 0.6317$ $n_1 = 12$

$\bar{y}_2 = 40.104$ $s_2 = 0.5314$ $n_2 = 8$

$\bar{y}_3 = 38.892$ $s_3 = 0.5108$ $n_3 = 10$

$H_0: \mu_1 = \mu_2$
$s_p = 0.5947$
$t_{18} = -10.28$, $p < 0.0005$

$H_0: \mu_1 = \mu_3$
$s_p = 0.5804$
$t_{20} = -6.35$, p-value < 0.0005

$H_0: \mu_2 = \mu_3$
$s_p = 0.5199$
$t_{16} = 4.91$, p-value < 0.0005

8.11a $H_0: \mu_D = 0$
Normality check ($p = 0.001$).
$\bar{d} = -1.125$, $s_d = 0.718$, $n = 4$
$t_3 = -3.13$, $0.05 < p < 0.1$
$0.10 > p > 0.05$

8.11b The sample size should be increased. There is also a problem with normality so a different method may need to be used.

Chapter 9: Relations

9.1

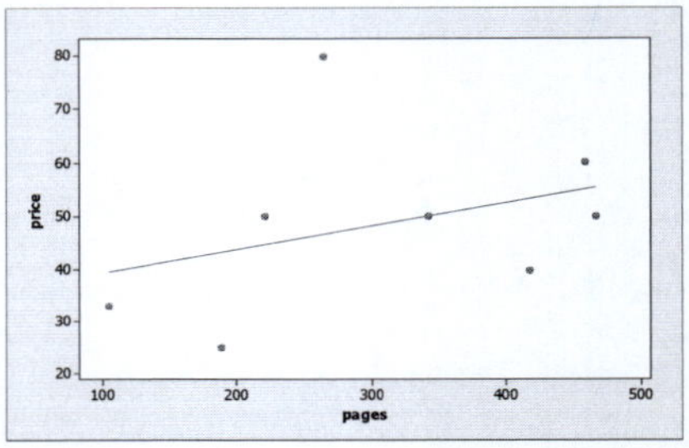

The scatter plot does not indicate a strong linear relationship between pages and price. There is one outlier – a book with 264 pages and a price of $79.95. This outlier may violate the assumptions that the residuals have a normal distribution and a constant standard deviation.
$\widehat{price} = 34.97 + 0.0438\,pages$

$H_o: \beta = 0$
$t_6 = 0.91$, $p > 0.5$

9.2

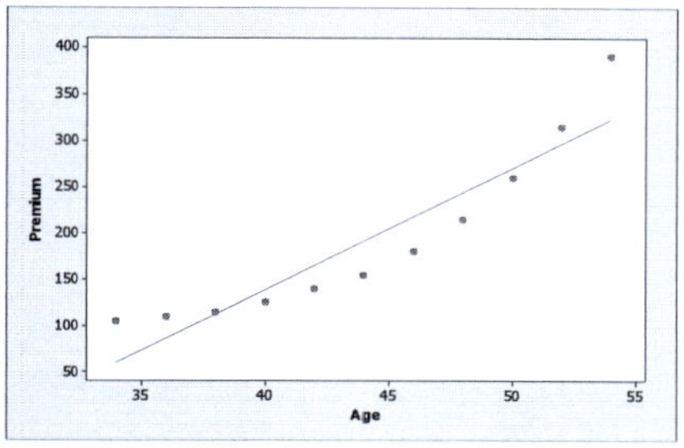

We have chosen male non-smoker.
The scatter plot shows some curvature, indicating that a straight line may not be the best model for these data.
The regression line is
$premium = -389 + 13.2age$
$r^2 = 0.866$. Thus 86.6% of the variation in premium is explained by age. This is a reasonable fit.
60 is outside the range so should not be used for prediction.
Note that the assumption of linearity has not been satisfied.

9.3

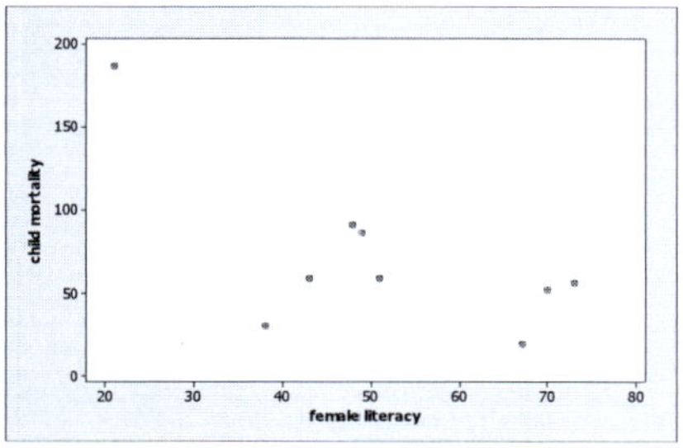

The scatter plot shows that linear regression is not appropriate for these data:

- there is one outlier (Yemen).
- there is no linear relation in the remaining data.

The missing values should not be estimated from these data:

- any regression line calculated using the data would be unreliable because of the violation of the assumptions;
- the missing values are of female literacy, which is the independent variable. The independent variable cannot be estimated from the dependent variable.

9.4a All subjects:

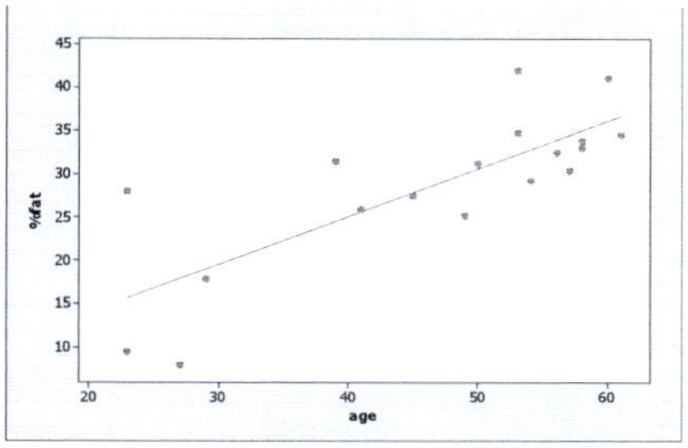

The LSR equation is
$\%fat = 3.22 + 0.548age$
H_0: $\beta = 0$
The assumptions of linearity, equal spread and normality appear to be satisfied. (Plot residuals to check)
$t_{16} = 5.17$, p-value < 0.0005
As age increases by one year, body fat percentage will increase by about half of one percent on average.

9.4b Females only:

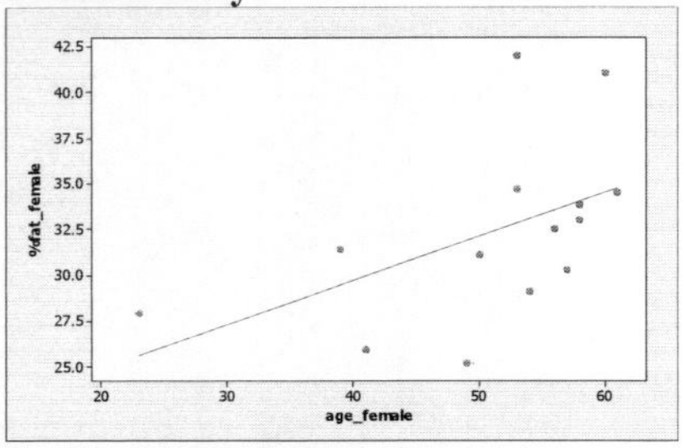

$\%fat = 20.1 + 0.24 age$

The assumptions of linearity, equal spread and normality appear to be satisfied. (Maybe some curvature)

H_0: $\beta = 0$

$t_{12} = 2.03$, $0.05 <$ p-value < 0.1

A different relation between %fat and age exists for the two groups so the data should not be combined.

9.5

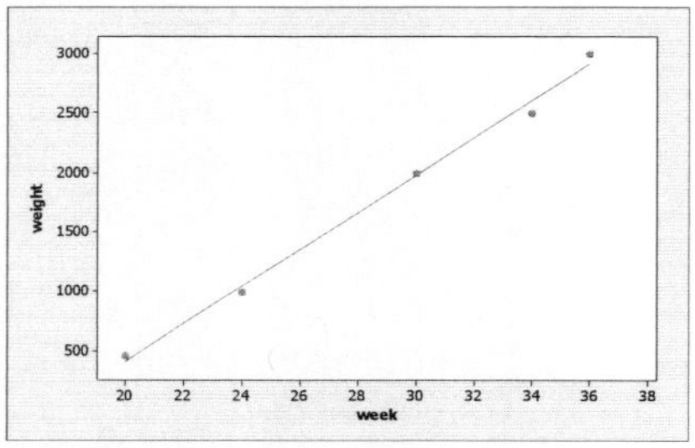

The LSR equation is

$weight = -2716 + 156.4 week$

The assumptions of linearity, equal spread and normality are all OK.

H_0: $\beta = 0$

$t_3 = 23.6$, p-value < 0.0005

As length of gestation increases by 1 week, foetal weight increases by 156 grams on average.

If birth occurs at 40 weeks, it is not reasonable to use this model to predict birth weight since a value of 40 is outside the range of x values. For this reason, it would not be reasonable to give a confidence interval for birth weight.

9.6

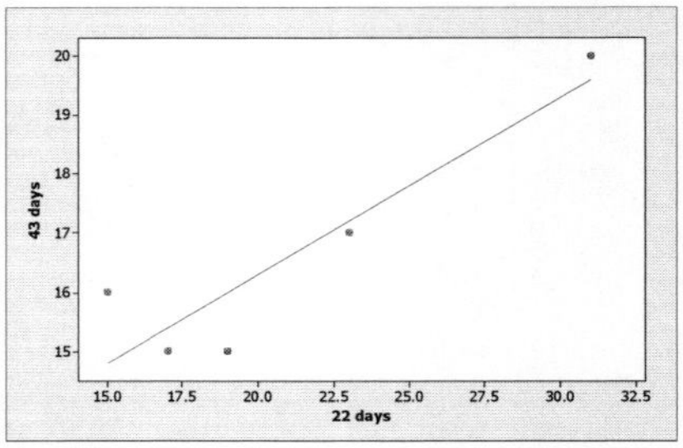

9.6a A positive relation exists.

9.6b The LSR equation is

$days43 = 10.3 + 0.3 days22$

r = 0.915 There is a strong positive linear relation.

9.6c H_0: $\beta = 0$

The assumption of linearity is suspect.

$t_3 = 3.93$, $0.02 <$ p-value < 0.05

For each extra percentage in porosity at 22 days we would expect the porosity to be 0.3% higher at 43 days.

9.6d

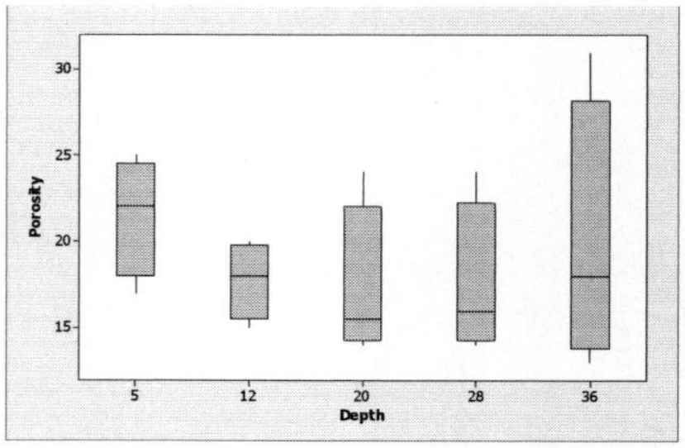

The comparative box plots show variability in both the spreads and the medians for the 5 different depths.

9.6e

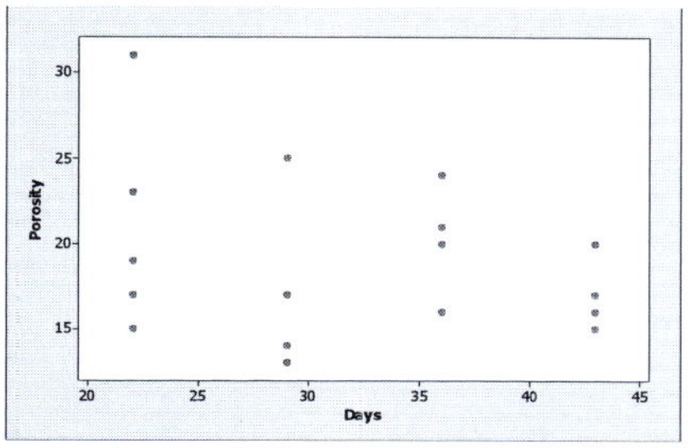

There appears to be a very weak negative relation with lots of scatter. The relation is not statistically significant.

H_0: $\beta = 0$

$t_{18} = -0.94$, $0.2 < \text{p-value} < 0.5$

9.7a $r = 0.69$

9.7b $r^2 = 0.472$

Thus, 47.2% of the variation in scores for boys is explained by scores for girls.

9.7c

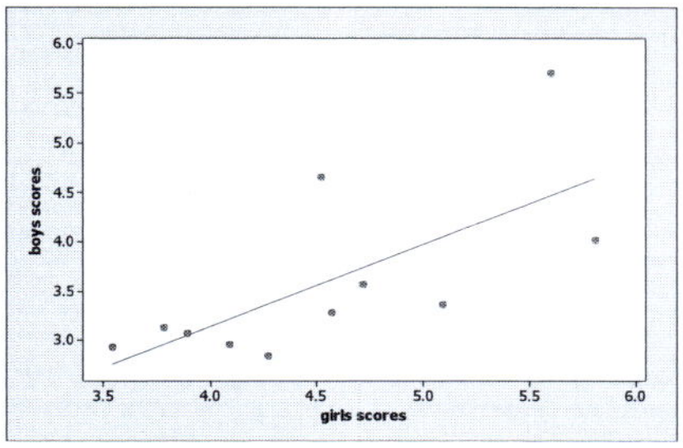

9.7d

The LSR equation is:

$\hat{y} = -0.190 + 0.833x$

9.7e

If boys' scores are used as the independent variable instead of girls' scores, the equation of the line will be different but the values of the correlation coefficient and goodness-of-fit statistics would remain the same.

From the information available, it is impossible to determine whether the boys' scores or girls' scores should be considered as the independent variable.

9.8

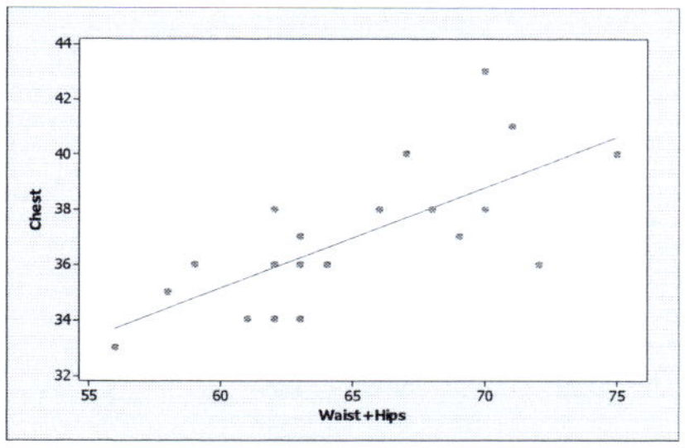

This scatter plot suggests that a straight line would be a suitable model.

The LSR equation is

$\widehat{Chest} = 13.22 + 0.37\,WaistHips$

H_0: $\beta = 0$

The assumptions of linearity, equal spread and normality appear to be satisfied. (Plot residuals to check)

$t_{18} = 4.38$, $\text{p-value} < 0.0005$

9.9

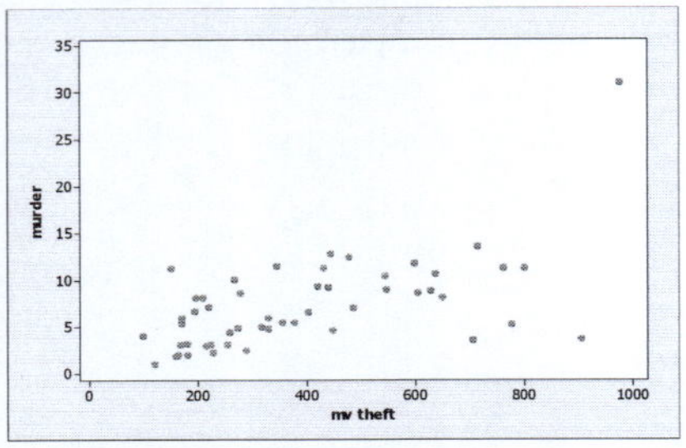

If the value for D.C is not omitted from the analysis, the normality assumption will be violated.

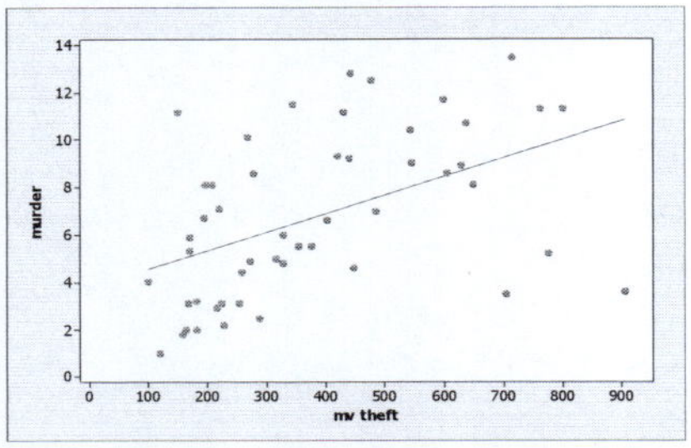

After removal of the outlier, the LSR equation is

$Murder = 3.8024 + 0.0078 MvTheft$

H_0: $\beta = 0$

The assumptions of linearity, equal spread and normality (check with plot of residuals) are now satisfied.

$t_{48} = 3.69$, p-value < 0.0005

For each extra 1,000 in the motor vehicle theft rate, the murder rate will by higher by 8, on average.

9.10

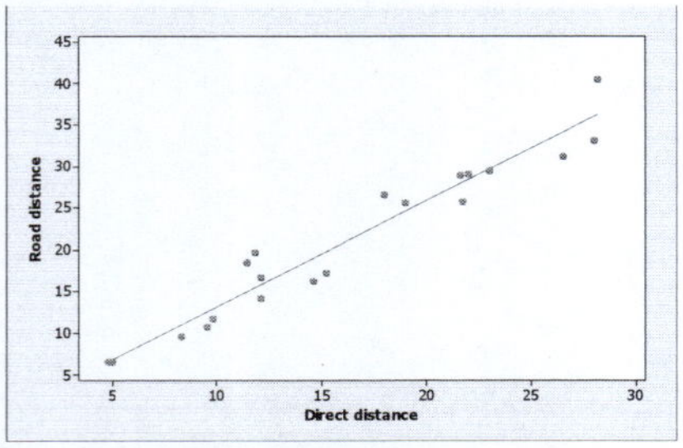

The regression equation is

$Road = 0.3791 + 1.2694 Direct$

H_0: $\beta = 0$

The assumptions of linearity, equal spread and normality appear to be satisfied.

$t_{18} = 16.66$, p-value < 0.0005

The goodness-of-fit is 93.9%, which indicates a very good fit.

Chapter 10: Cross-tabulations

10.1 H_0: $\pi_0 = 0.7$, $\pi_1 = 0.2$, $\pi_n = 0.1$

All expected counts are ≥ 5

$\chi_2^2 = 6.3$, $0.02 <$ p-value < 0.05

10.2 H_0: there is no association between spinal anaesthesia and colic.

All expected counts are ≥ 5

$\chi_1^2 = 6.66$, $0.005 <$ p-value < 0.01

10.3 All expected counts are ≥ 5

w = 1.4, 95% CI is (0.437, 4.485)

No association is apparent since the 95% CI for ω contains 1.

$\chi_1^2 = 0.32$, p-value > 0.5

10.4 **Make own breakfast:**

Not all expected counts are ≥ 5 so the results should be treated with caution.

w = 0.214, 95% CI is (0.022, 2.091)

No association is apparent since the 95% CI for ω contains 1.

Pick up dirty clothes:

Not all expected counts are ≥ 5 so the results should be treated with caution.

w = 0.633, 95%C I is (0.095, 4.212)

No association is apparent since the 95% CI for ω contains 1.

Wash car:

All expected counts are ≥ 5

w = 1.786, 95% CI is (0.523, 6.104)

No association is apparent since the 95% CI for ω contains 1.

10.5a

H_o: the proportion of suspect transactions has not changed between years.

All expected counts are ≥ 5

$\chi_1^2 = 12.03$, $0.0005 < \text{p-value} < 0.001$

10.5b

H_o: type of cash dealing and type of transaction are not associated.

All expected counts are ≥ 5

$\chi_2^2 = 127.1$, $\text{p-value} < 0.0005$

10.6a

H_0: $\pi_{90} = 2/3$, $\pi_{92} = 1/3$

All expected counts are ≥ 5

$\chi_1^2 = 1.38$, $0.2 < \text{p-value} < 0.5$

The proportion dieing in 1992 could be half that in 1990.

10.6b

Suppose that the population at risk is 10000:

w = 2.367, 95% CI is (1.817, 3.083)

since the 95% CI for ω does not contain 1, there is an association and the proportions are different for 1990 and 1992.

10.7 H_0: there is no association between scientific background and opinion.

Note: one expected count is only 4.5 but this is probably OK.

$\chi_4^2 = 13.91$, $0.005 < \text{p-value} < 0.01$

10.8 H_0: $\pi_l = 0.35$, $\pi_m = 0.27$, $\pi_h = 0.38$

All expected counts are ≥ 5

$\chi_2^2 = 4.6$, $0.05 < \text{p-value} < 0.1$

10.9a

H_0: $\pi_1 = \pi_2 = \pi_3 = 1/3$

All expected counts are ≥ 5

$\chi_2^2 = 1.6$, $0.2 < \text{p-value} < 0.5$

10.9b

H_0: work attachment is independent of work status.

All expected counts are ≥ 5

$\chi_6^2 = 134.8$, $\text{p-value} < 0.0005$

10.9c

OR: Work status/attachment

PT,FT/mod,high: w = 10.8

95% CI for ω is (4.06, 28.72)

PT,FT/low,high: w = 4.5

95% CI for ω is (0.66, 30.54)

NE,FT/mod,high: w = 9.9

95% CI for ω is (1.72, 56.95)

NE,FT/low,high: w = 354.75

95% CI for ω is (47.45, 2652)

LW,FT/mod,high: w = 14.85

95% CI for ω is (2.75, 80.29)

LW,FT/low,high: w = 7.25

95% CI for ω is (9.15, 602.7)

10.10a

H_0: no association between age of mother at birth of first child and breast cancer status.

All expected counts are ≥ 5

$\chi_1^2 = 1.77$, $0.1 < \text{p-value} < 0.2$

10.10b

w = 0.75

95% CI for ω is (0.487, 1.155)

10.10c

No association is apparent between age of mother at birth of first child and breast cancer status.

10.11

H_0: $\pi_{tc} = 9/16$, $\pi_{tp} = 3/16$, $\pi_{dc} = 3/16$, $\pi_{pc} = 1/16$

All expected counts are ≥ 5

$\chi_3^2 = 1.47 \text{p-value} > 0.5$

One of $n\pi$ and $n(1-\pi) < 5$. So, CLT does not apply and the results should be treated with caution.

Index